MANAGING BASIN INTERDEPENDENCIES IN A HETEROGENEOUS, HIGHLY UTILIZED AND DATA SCARCE RIVER BASIN IN SEMI-ARID AFRICA

MANAGING BASIN INTERDEPENDENCIES IN A HETEROGENEOUS, HIGHLY UTILIZED AND DATA SCARCE RIVER BASIN IN SEMI-ARID AFRICA

The case of the Pangani River Basin, Eastern Africa

DISSERTATION

Submitted in fulfilment of the requirements of
the Board for Doctorates of Delft University of Technology
and of the Academic Board of the UNESCO-IHE Institute for Water Education
for the Degree of DOCTOR
to be defended in public
on Thursday, 13 October 2016 at 15.00 hours
in Delft, the Netherlands

by

Jeremiah Kipkulei KIPTALA
Master of Science in Water Management,
UNESCO-IHE Institute for Water Education, the Netherlands
born in Baringo, Kenya

This dissertation has been approved by the
promotor: Prof.dr.ir. P. van der Zaag
copromotor: Dr. Y.A. Mohamed

CRC Press/Balkema is an imprint of the Taylor & Francis Group, an informa business

Published by:
CRC Press/Balkema
PO Box 11320, 2301 EH Leiden, the Netherlands
Pub.NL@taylorandfrancis.com
www.crcpress.com – www.taylorandfrancis.com
ISBN: 978-1-138-03609-3

ABSTRACT

The concept of integrated water resources management (IWRM) aims to integrate all relevant elements of water resources in a comprehensive and holistic way. The combined management of blue and green water resources in a river basin and their spatial and temporal distribution have to be considered for an IWRM plan. Green and blue water follow distinct pathways and are associated with different water use practices. In sub-Saharan Africa, rainfed and supplementary irrigated crops – relying mainly on green water resources - dominate upper landscapes. In downstream areas, blue water uses are often confined along the river channels, mainly for hydropower and environment. Over time, and due to population growth and increased demands for food and energy, water demands for both green and blue water resources have increased. Often the increasing green water use in upstream catchments has led to declining blue water resources in the downstream parts. The classical water resources management approach often focusses on the blue (runoff) water resources only. This is attributed to limited information on the temporal and spatial distribution of the green water (soil moisture) in the basin. Obviously, this has hampered the development of sound and sustainable IWRM plans in those basins. Therefore, an integrated analytical system for the entire river basin, incorporating both green and blue water resources is needed to assess upstream-downstream interdependencies, and to provide boundary conditions for an optimal water management plan at river basin scale.

The management of basin interdependencies - particularly in cases of water scarcity - fundamentally depends on available knowledge and data. This thesis applied various approaches – some of which being innovations - to generate locally validated information in a heterogeneous, highly utilized and data scarce river basin in Africa - the Pangani river basin. An accurate assessment of (i) water availability, (ii) water use, (iii) water productivity, and (iv) water value, allowed the identification of basin interdependencies, and the quantification of tradeoffs and synergies between different green and blue water uses.

The upper Pangani River Basin can be considered a closed basin due to intensive water use, mainly for agriculture. The many irrigation systems developed by smallholder farmers consist of complex and intricate networks of earthen canals and provide supplementary irrigation to otherwise rainfed crops, combining water from precipitation (green water) and river abstractions (blue water). There is very little official information about water use and water productivity of these gravity-fed irrigation systems. The increasing water uses for irrigation upstream has generated externalities and water related conflicts among various users in the basin. With time, the environment has also been affected as most of the perennial tributaries have become seasonal.

In semi-arid areas such as the Upper Pangani River Basin, evaporative water use constitutes the largest component of the hydrological cycle, with runoff barely exceeding 10%. The evaporative flux is a function of land cover and land management practices. Information of the spatially distributed land use and land cover (LULC) is required to (i) assess green water use per LULC, and (ii) characterize hydrological model parameters that provide the link between green water use and blue water flows. Through remote sensing analysis, sixteen different LULC types were identified and classified using their unique temporal phenological signatures. The methodology relied on freely available satellite data on vegetation provided by the Moderate-resolution Imaging Spectroradiometer (MODIS). The data has 8 day temporal and 250 m spatial resolution, and covers the hydrological years of 2009 to 2010. Unsupervised and supervised clustering techniques were utilized to identify various LULC types with aid of ground data obtained during two rainfall seasons (short and long) in the river basin. The multi-temporal MODIS data and long time series ensured correct timing of change events in the vegetation growth. The overall classification accuracy was 85%, with producer's accuracy of 83% and user's accuracy of 86% (at 98% confidence level). The individual classes showed relatively good accuracies of over 70%, except for barelands. Lower accuracies were observed for the smaller LULC classes. This uncertainty was attributed to the moderate resolution of MODIS gridded data (250-m). The inaccuracies were corrected using the Kappa statistic (K). The derived LULC classes were consistent with the FAO-SYS land suitability classification. Additional checks were made against local databases of smallholder irrigation and large scale irrigation (sugarcane cultivation), and the results showed close agreements (74% and 95%, respectively), with a fairly good geographical distribution.

Accurate estimation of actual evapotranspiration (ET) for the 16 different LULC types in a data scarce region is challenging. This study used the MODIS satellite data and Surface Energy Balance Algorithm of Land (SEBAL) to estimate the actual ET for 138 images, with 250-m, and 8-day resolution for the period 2008 to 2010. A good agreement was attained for the SEBAL ET against various validations. The estimated ET (open water) for Nyumba ya Mungu (NyM) reservoir showed a good correlations against pan evaporation data ($R^2 = 0.91$; Root Mean Square Error (RMSE) of less than 5%). An absolute relative error of 2% was calculated based on the mean annual water balance estimates of the reservoir. The estimated ET for agricultural land use classes indicated a consistent pattern with the seasonal variability of the crop coefficient (K_c) based on the Penman-Monteith equation. The ET estimates for the mountainous areas were significantly suppressed at higher elevations (above 2,300 masl), which is consistent with the reduced potential evaporation in those areas. The ET estimates were comparable to the global MODIS 16 ET data in variance (significant at 95% confidence) but not with respect to the mean. This level of significance provides optimism but caution in the use of the freely available global ET datasets that have not been locally validated.

A major limitation in deriving remote-sensed ET especially for land use types at higher elevations in the humid to sub-humid tropics is the persistent cloud cover. Those clouded pixels were corrected by interpolation based on the next and/or previous images. Although, the cloud filling procedure benefited from the multispectral set of MODIS images, it still may introduce uncertainties in the final results. For the whole basin the estimated ET accounted for 94% of the total precipitation with an outflow closure difference of 12% to the measured discharge at the outlet. The bias (12%) was within the uncertainty range (13%) at 95% confidence level. The water balance analysis clearly showed that the basin is fastly closing. Therefore, it is important and timely to improve water productivity through improved water efficiency and water re-allocation in the Upper Pangani basin.

Quantifying the hydrological link between the spatially distributed green water use (evaporation) and blue water (river flows) is essential for assessing interdependencies at the basin scale, though it is challenging. Physically based spatially distributed models are often used. But these models require enormous amounts of data, which may result in equifinality, and hence make such models less suitable for scenario analyses. Furthermore, these models often focus on natural processes and fail to account for anthropogenic influences. This study adopted an innovative methodology for quantifying blue and green water flows. The methodology uses ET and soil moisture derived from remote sensing as input data to the Spatial Tools for River basin Environmental Analysis and Management (STREAM) model. To cater for the extensive irrigation water abstractions, an additional blue water component (Q_b) was incorporated in the STREAM model to quantify irrigation water use. To support model parameter identification and calibration, two hydrological landscapes (wetlands and hill-slope) were identified using field data and topographical maps. The model was calibrated against discharge data from five gauging stations and showed a good performance especially in the simulation of low flows. The Nash-Sutcliffe Efficiency of the natural logarithm (E_{ns_ln}) of discharge were greater than 0.6 in both calibration and validation periods. At the outlet gauging station, the E_{ns_ln} coefficient was even higher (0.90). The only challenge in using remotely sensed data (8-day) as input in hydrological models are in processes such as interception that have time scales of less than 8 days. Such hydrological processes have to be calculated outside the model thus introducing additional uncertainties.

During low flows, Q_b consumed nearly 50% of the river flow in the Upper Pangani basin. Q_b for irrigation was comparable to the field based net irrigation estimates with less than 20% difference. A number of water management scenarios on water saving and impacts of increased water use were explored. The modified STREAM model showed a potential to be replicated in other landscapes with complex interactions between green and blue water uses. The model flexibility offers the opportunity for continuous model improvement when more data becomes available. The output from the model, mainly the information on green-blue water flows, was used as input in the water productivity analysis.

Although water productivity is a key indicator in basin water resources management, it is not readily available, in particular for natural landscapes. The measures to improve water productivity are also unique to different river basins. This study computed water productivity in the Upper Pangani basin using a combination of remote sensing models. The models were based on the Monteith's framework for dry matter production to estimate above-ground biomass production in agricultural and natural landscapes. SEBAL algorithm was used to compute biomass production from

MODIS images. The gridded biomass production was then converted to crop yield, and amount of carbon sequestered. These were then converted to gross returns using their market prices. This study included gross returns from carbon credits and other ecosystem services in the concept of economic water productivity (EWP). The EWP showed the levels of water use and when formulated as production functions it can show the scope for improvements and provide for a trade-off analysis in a river basin. The biophysical productivity (biomass and crop yield) and water yields also provided insights into the water value society attaches to certain natural land use activities.

Irrigated sugarcane and rice achieved the highest water productivities both in biophysical and economic values – well within the ranges reported in the literature. However, the productivities of rainfed and supplementary irrigated banana and maize showed a wide spatial variability, and were significantly lower than potential. The supplementary irrigated crops that combine green and blue water, however, achieved a higher economic productivity of blue water than fully irrigated crops. In situations of water scarcity, it is therefore prudent to allocate water resources to supplementary irrigated crops rather than to fully irrigated crops. This thesis developed explicit analytical relationships between biomass production and ET for irrigated, rainfed and natural landscapes for the Pangani River Basin. These relationships, which were formulated as production functions, showed the potential of improving the productivity of rainfed and supplementary irrigated agriculture in the basin. The frequency distribution of biomass production at pixel scale provided additional evidence for improvements in water productivity.

An integrated hydro-economic model (IHEM) was developed in order to integrate green and blue water resources, for multi-objective analysis of water uses in the entire Pangani River Basin. The IHEM, which aims to optimize blue water use, was formulated innovatively to account for the full water balance. This has been done by incorporating the green water resources through their production functions in the Upper Pangani Basin. The analysis focuses on three primary objective functions: i) hydropower production, ii) fully irrigated agriculture, where crop water requirements were met by blue water, and iii) supplementary irrigation, where crop requirements were met by both green and blue water. The analysis also considered five socio-environmental objectives that were derived from key stakeholders and expert knowledge. The results showed that agricultural water use (supplementary and fully irrigated) achieves relatively high water productivities and competes with hydropower, urban water use and the environment. Firm energy (provided at 90% reliability) favours constant moderate flow conditions throughout the year, which then competes with the environment that requires both high and low flow conditions, depending on the season. This study showed that improving rainfed maize through supplementary irrigation has a slightly higher marginal water value than fully irrigated sugarcane. For achieving sustainability of the river basin, agricultural water use should be balanced with other economic, social and environmental water requirements. Because water demand for hydropower is largely non-consumptive, hydropower production can, in theory at least, be seasonalized for conjunctive water use with the environment.

The IHEM model provided the blue water balance of the Lower Pangani Basin and showed that the Upper Pangani River Basin contributes 82% of total blue water. Evaporation from NyM reservoir constitutes about 28% of total inflows into the reservoir. The water use at the Kirua swamp, though constrained by water regulation at the NyM reservoir, is equivalent to US$ 8 million per year of potential hydropower revenue. The study showed that the minimum environmental flow for the Pangani estuary is guaranteed by the flow requirement from hydropower production in the two hydro-electric plants located near the outlet. Furthermore, the high flow requirement for the estuary is presently sustained by the unregulated flows from Mkomazi and Luengera tributary rivers. The scenario analyses showed various levels of trade-off between competing water users. Any measure that increases inflows into the reservoir or reduces water demands downstream of the reservoir would result in an operating policy that minimizes reservoir evaporation and provides more naturalized outflows downstream. Investment in interventions to reduce non-productive soil evaporation from irrigated mixed crops in upstream catchments resulted in increased blue water inflows into NyM reservoir that would increase hydropower revenue by US$ 2 million per year. This is equivalent to 33 US$ ha^{-1} yr^{-1} which could be available for investments in soil and water conservation, a potential for payment for environmental services (PES). The increase in revenue is in addition to un-quantified ecosystem services that would result from increased river flows downstream.

Although this study could clearly demonstrate the advantages of integrated hydroeconomic modelling by including green water use upstream and blue water use downstream, deriving an accurate water value for the ecosystem services, in particular for wetlands, proved a challenge. The environmental values can be incorporated into the non - economic production functions (used as constraints in our model) to provide a wider variety of options and trade-offs for stakeholders and decision-makers.

x

ACKNOWLEDGEMENT

Since the start of the PhD research in 2010, I have received enormous support, guidance and advice from my supervisors: Prof Pieter van der Zaag, Dr. Yasir Mohamed and Dr. Marloes Mul. I remember the unprecedented effort, critical and innovative insights each one put towards this scientific research. I also remember the field visits and the encouragements you provided through all periods of the research.

The study could not have been possible without the funding from the Netherlands Ministry of Development Cooperation (DGIS) through the UNESCO-IHE Partnership Research Fund (UPaRF). The study was carried out in the framework of the Research Project 'Upscaling small-scale land and water system innovations in dryland agro-ecosystems for sustainability and livelihood improvements' (SSI-2). I am gratefully and acknowledge data and information provided by the following organizations: Pangani Basin Water Office & IUCN (Moshi, Tanzania), Irrigation Department in the Ministry of Water and Irrigation (Moshi, Tanzania), Tanzania Meteorological Agency (Dar es Salaam, Tanzania), TANESCO (Hale, Tanzania) and Kenya Meteorological Department (Nairobi, Kenya).

I thank my PhD colleagues for the discussions and time (including the fun) that made life easier during the study, in particular Dr. Hans Komakech, Ceaser Orup, Micah Mukolwe, Peter Matuku, Adoko Kapko and Dr. Frank Masese. I also acknowledge the many discussions with researchers from Smallholder Systems Innovations (SSI) partner institutions that went into this research particularly: Dr. Tumaini Kimaru (late), Dr. Victor Kongo and Dr. Deogratias Mulungu. I gratefully acknowledge the MSc students within the SSI that made contributions to this study and specifically Magreth Mziray, Edmund Musharani, Benson Bashange and William Senkondo. Much appreciation also goes to the staff at the Department of Integrated Water Systems & Governance (IWSG) of UNESCO-IHE, in particular the support offered by Ms. Susan Graas, Dr. Ilyas Masih, Dr. Jeltsje Kemerink as well as Prof. Amaury Tilmant. I also acknowledge the support from Dr. Muhammad Jehanzed Masud of the University of Agriculture, Faisalabad and my colleagues at the Department of Civil Engineering, Jomo Kenyatta University of Agriculture and Technology, Nairobi, Kenya, in particular Prof. Eng. Geoffrey Manguriu, Dr. John Mwangi, Ms. Purity Kibetu and Eng. Simon Mdondo.

I would thank the editors and the anonymous reviewers who provided valuable comments and suggestion in the publications emanating from this PhD research.

Lastly, I would most thank my family and friends for supporting and allowing me to be away during part of the study, specially, my wife Ms. Josephine Chebet, my daughter Stacey Jeptum and my sons Andries Chesaro and Amaury Chesang. I pay tribute to my late father Mzee Alfred Komen for the sacrifices he made for my earlier education that has culminated in this thesis.

TABLE OF CONTENT

LIST OF SYMBOLS

Symbol	Parameter description	Value	Dimension/Unit
B	Biomass		Kg ha^{-1} yr^{-1}
C	Capillary rise		L^3 T^{-1}
C_{max}	Maximum Capillary rise		L^3 T^{-1}
C_{min}	Minimum Capillary rise		L^3 T^{-1}
CV	Coefficient of Variation		-
c_r	Separation coefficient for net precipitation		-
D	Threshold value for interception		L
E	Total Evaporation		L T^{-1}
E_{ns}	Nash-Sutcliffe coefficient		-
E_{ns_ln}	Nash-Sutcliffe coefficient (natural logarithm)		-
f	Soil moisture depletion fraction		-
G	Soil heat flux		W m^{-2}
H	Sensible heat flux		W m^{-2}
H_o	Elevation above nearest open water		L
H_s	Normalized DEM above H_o		L
$HAND$	Height Above the Nearest Drainage		L
I	Interception		L T^{-1}
T	Transpiration		L T^{-1}
E_o	Open water evaporation		L T^{-1}
$E_{o(b)}$	Open water evaporation from water balance		L T^{-1}
$E_{o(p)}$	Open water evaporation from pan measurements		L T^{-1}
E_p	Pan evaporation		L T^{-1}
E_s	Evaporation from the soil		L T^{-1}
ET	Actual Evaporation and Transpiration		L T^{-1}
G	Heat flux density into the water body		M T^{-1}
g	Gravity	9.81	m s^{-1}
λ	Latent heat coefficient	$2{,}47 \times 10^6$	J kg^{-1}
Λ	Evaporation fraction		-
λE	Instantaneous latent heat flux		W m^{-2}
ρ_ω	Density of water	1000	kg m^{-3}

K	Kappa statistic		-
K_p	Pan coefficient factor	0.81	-
K_c	Crop coefficient		-
K_o	Time scale, overland flow		T
K_q	Time scale, quick flow		T
K_s	Time scale, slow flow		T
moi	Moisture content		-
P	Precipitation		L T^{-1}
P_e	Net precipitation		L T^{-1}
P_c	Production costs		US$
P_g	Gross farm gate price		US$
P_n	Net farm gate price		US$
Q_b	Blue water use		L^3 T^{-1}
Q_d	River net abstractions		L^3 T^{-1}
Q_g	Green water use		L^3 T^{-1}
Q_o	Observed discharge		L^3 T^{-1}
Q_{of}	Overland flow		L^3 T^{-1}
Q_{qf}	Quick flow		L^3 T^{-1}
Q_{sf}	Slow flow		L^3 T^{-1}
Q_s	Simulated discharge		L^3 T^{-1}
Q_u	Excess overflow from unsaturated		LT^{-1}
R	Correlation coefficient		-
R^2	Coefficient of determination		-
R_n	Net Radiation		W m^{-2}
S	Storage, volume		L^3
S_b	Blue water storage		L
S_u	Storage unsaturated zone		L
$S_{u,max}$	Maximum storage of unsaturated zone		L
$S_{u,min}$	Minimum storage of unsaturated zone		L
S_s	Storage saturated zone		L
$S_{s,,q}$	Threshold value for quick runoff		L
$S_{s,max}$	Threshold value for direct runoff		L
$S_{s,min}$	Threshold value for slow flow		L
$S_{c,min}$	Threshold value for capillary rise		L
W	Atmospheric moisture content		Kg m^{-2}
dS/dt	Change in storage of the catchment		L^3 T^{-1}

LIST OF ACRONYMS

ALSE	Agricultural Land Suitability Evaluator
AOI	Area Of Interest
APAR	Absorbed Photosynthetic Active Radiation
AVHRR	Advance Very High Resolution Radiometer
DEM	Digital Elevation Model
E-Pan	Evaporation Pan
EWP	Economic Water Productivity
EOS	Earth Observation System
EWURA	Energy and Water Utilities regulatory Authority
FAO	Food and Agriculture Organization
FEWS	Famine Early Warning Systems
GAMS	General Algebraic Modelling System
GEPIC	GIS – based Environmental Policy Integrated Climate model
GIS	Geographic Information System
GLC	Global Land Cover
GLDAS	Global Land Data Assimilation System
GMAO	Global Modelling and Assimilation office
GRACE	Gravity Recovery And Climate Experiment
HEP	Hydro-Electric Power
IHEM	Integrated Hydro-Economic Model
IFRI	International Food Policy Research Institute
IPP	Independent Power Producers
ISODATA	Iterative Self Organizing Data Analysis Technique
IUCN	International Union for Conservation of Nature
IWRM	Integrated Water Resources Management
LAI	Leaf Area Index
LST	Land Surface Temperature
LPDAAC	Land Processes Distributed Active Archive Center
LULC	Land Use and Land Cover
MAE	Mean Absolute Error

m.a.s.l	Meters above sea level
MODIS	MODerate resolution Imaging Spectrometer
MSG	Meteosat Second Generation
NASA	National Aeronautics Space Administration
NDVI	Normalized Difference Vegetation Index
NOAA	National Oceanic and Atmospheric Administration
NPF	New Pangani Falls hydro-electric plant
NyM	Nyumba ya Munyu reservoir
PES	Payment for Environmental Services
PBWO	Pangani Basin Water Office
RMSE	Root Mean Square Error
RS	Remote Sensing
SASRI	South Africa Sugar Research Institute
S-SEBI	Simplified Surface Energy Balance Index
SEBAL	Surface Energy Balance Algorithm for Land
SEBS	Surface Energy Balance System
SIs	System Innovations
SSI	Smallholder systems innovations in Integrated Watershed Management
STDEV	Standard Deviation
STREAM	Spatial Tools for River basin Environmental Analysis and Management
STRM	Shuttle Radar Topography Mission
SWAT	Soil and Water Assessment Tool
TANESCO	Tanzania Electric Supply Company Limited
TRMM	Tropical Rainfall Measuring Mission
TSEB	Two Source Energy Balance
TPC	Tanzania Plantation Company
UNEP	United Nations Environment Programme
USGS	United State Geographical Survey
VI	Vegetation Index
WCD	World Commission of Dams
WP	Water Productivity

Chapter 1

INTRODUCTION

1.1 WATER MANAGEMENT ISSUES AND CHALLENGES

Water is an important natural resource to all forms of life and existence, and it forms the backbone for economic productivity and social wellbeing. This fundamental role and the growing demand amongst various users is becoming a great challenge to water resource managers at river basin scale. Water demand already exceeds supply in many parts of the world, and as population continues to rise, and economies grow, more areas are expected to experience water scarcity (Vörösmarty and Sahagian, 2000; Smakhtin et al., 2004; Bos et al., 2005; Gourbesville, 2008). Water managers are also facing a massive challenge as they seek to balance human water demand with ecological needs. More water is also required to fulfil increasing energy demands from hydropower and biofuels (de Fraiture et al., 2008). This, according to Perry (1999), promotes an approach that links sources, uses, losses, and reuses by different land-use categories and environmental systems present within river basins.

The situation in Africa becomes even more pronounced as over 60% of the total population relies on water resources that are limited and highly variable (UNEP, 2010). 75% of the continents' cropland is located in arid and semi-arid areas, with high variability of hydro-climatic condition, where irrigation can greatly improve productivity and reduce poverty (Smith, 2004; Vörösmarty et al., 2005). Only 4.8% of global hydropower potential is exploited (Gopalakrishnan, 2004). In sub Saharan Africa, 90% of agricultural land is rainfed with 70 – 90% of the exploited 'blue water' being used for irrigation (Rockström, 2000). Moreover, environmental values and its ecological benefits to livelihoods of rural populations are being recognized and now the environment is accepted as a legitimate water user in river basins. All these sectoral water users are interdependent of each other and any measure to influence the productivity or allocation of one user will affect the productivity of the other user (Van der Zaag, 2007; 2010). This situation is expected to be exacerbated with future expected population increase, economic development and climate change.

Integrated water resources management based on the principle of economic productivity (efficiency) while ensuring equity and ecological integrity has potential to achieve conflicting and varying objectives of all water users in a river basin. Achieving economic productivity requires the understanding of the availability of water and a notion of how much of it will be needed, in what quantity, for how long, and for what purposes (Gürlük and Ward, 2009).

Since water availability is key to economic development, disputes over shared water resources continue to rise between different users/sectors. This is mainly because of the socio-economic difference and the physical linkages that exist especially between upstream and downstream users. We therefore need to recognize and institutionalize this upstream-downstream interdependence that may help build hydrosolidarity and cooperation among water users (Falkenmark and Folke, 2002; Van der Zaag, 2007).

The success of any dialogue or policy depends on the knowledge base, general trust in data sources and tools that will enable policy makers, planners and stakeholders to make well informed decision. It is along this vein that this research aims to provide basin-wide tools, and information to effectively manage basin interdependencies between different water users in the Pangani River Basin in Eastern Africa. The study is set in a heterogeneous, highly utilized river basin with distinct mountainous upper catchments generating most of the water resources and large grassland savannah in the lower catchments. Although this is a typical African catchment, the methodology, data requirements, and findings are generic and can be applied in any other region.

The Pangani basin which covers an area of 43,650 km^2 has an estimated population of 3.7 million people of which 90% lives in rural areas. 80% of the rural population depends, directly or indirectly on agriculture for their livelihood. Traditional irrigation systems are practiced by smallholders and several large scale farming enterprises also exist in the basin (IUCN, 2003; Komakech et al., 2010). The basin is also a major supplier of electricity from hydropower and hosts vital natural ecosystems such as mountain reserves, freshwater lakes, wetlands and the estuary. Detailed description and features of the river basin is provided in Chapter 2.

There is an increasing demand and competition for water resources. Agricultural interests are expanding with irrigation being adopted by many farmers to enhance productivity. There are also increasing water demands from urban water supply. Land use is changing as more forest land and natural vegetation is transformed into agricultural land. Between 1952 and 1982, Kilimanjaro's natural forest declined by over 41 km^2 and approximately 77% of the forest cover of the Pare and Usambara Mountains, the most densely populated areas of the Pangani River basin, has been lost to agriculture (IUCN, 2003). Hydropower production has declined due to reduced inflows into the reservoirs (IUCN, 2007). Environmental resources have also been affected by reduced river flows as far as the Pangani estuary, where salt intrusion is a problem (Sotthewes, 2008). Some farming and fisheries are thought to have declined also as a result of decreased fresh water flows in the Kirua swamp and the estuary (IUCN, 2003; Turpie et al., 2003). Furthermore, most perennial tributaries in the upper catchments of the Pangani River Basin have actually become seasonal in the last few decades.

According to Grossmann (2008), the Pangani River Basin can be classified as a 'closed basin', where all its available water has been used. Other studies have indicated that the basin is experiencing closure during periods of low flows especially in the lower parts of the basin, the Pangani estuary (PBWO/IUCN, 2009). With river basin closure, interdependencies increase and manifest themselves in alterations of the water cycle that create positive and negative externalities to different categories of users and the environment (Molle and Wester, 2009).

The Pangani Basin Water Office (PBWO) manages the water supply at the basin since 1991 following a new water policy. Water users apply for and are allocated water rights to certain amount of flows based on an understanding of supply of and demand for water in the basin. There are no defined mechanisms for allocation of water to different users and over 3,000 water rights have been issued by 2010. There are also different claims to water access rights and causes of water shortages in the basin (Komakech et al., 2010).

This situation and the increasing demand for water resource has generated water conflicts between various users. Sarmett et al. (2005) and Mbonile (2005) classify these conflicts as conflicts of scale (users of different sizes), conflicts of tenure (water rights) and conflicts of location (upstream and downstream users) depending on the power and position of the various users on the river basin (Box 1.1).

Box 1.1: Categories of Conflicts in Pangani Basin (Source: Sarmett et al., 2005; Mbonile, 2005).

- Conflicts of scale: Conflicts between users of different sizes and power in the basin. Such as the large scale plantations, using hundreds of litres of water per second through 'efficient' drip irrigation system, differ from small-scale users of traditional furrow systems with 'efficiency' as low as 14%.

- Conflicts of tenure: Tenure is the right to manage a resource. Small scale users in the basin are reluctant to apply and pay water rights, arguing that water is a 'gift' from God.

- Conflicts of location: Tanzania Electricity Supply Company (TANESCO) located downstream pays royalty to the Ministry of Water & Livestock Development for a 45 m^3 s^{-1} flow. Because of reduced rainfall and upstream abstractions, the company often receives as little as 15 m^3 s^{-1}, limiting hydropower production and creating national-level conflicts.

It is therefore important to plan water resources development, allocation, and management in a context of multiple uses of water based on the actual amount of water available in the basin, economic efficiency and with an understanding of the potential impacts (socio-economic and environmental). It is also important at this point to note that the Pangani River Basin is largely ungauged with limited hydro-meteorological data (Mul, 2009). This study therefore applied a methodology which uses freely available remotely sensed (RS) data to generate the information required for water resource planning in the river basin. The study also assessed the applicability and accuracy of using this RS data.

Fig. 1.1 presents a problem tree that summarizes the causes and interdependencies of the dominating issues based on previous studies done on the Pangani basin under the SSI-1 Programme (SSI, 2009) and PBWO/IUCN (IUCN, 2003; IUCN, 2007, PBWO/IUCN, 2009). The main problem of the river basin can be summarized as growing water scarcity and lack of a clear-cut water allocation policy that is manifested in sub-optimal water use and conflicts between various water users in the river basin. The resilience of most farming systems is low due to the large variability of the hydro-climatic conditions and a limited capacity to adapt. This often results in low crop yields, on average below 1 ton per ha for smallholders (Makurira et al., 2010). Population growth and increasing food production to meet not only the local but also global food demands imposes high pressures on the (limited) water resources. It is also exerting pressure on the traditional farming practices because of reduced farm sizes. Degradation of natural environments such as forests, riparian vegetation and wetlands has occurred in the last few decades. These natural systems provide a wealth of ecosystems goods and services especially to local communities (Costanza et al., 1997; de Groot et al., 2012). The vulnerability of the poor population who rely on the ecosystem services has also increased (Malley et al., 2007; Enfors and Gordon, 2007; 2008).

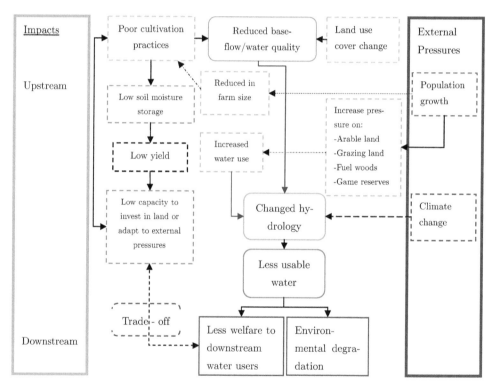

Fig. 1.1: Cause and effect tree of dominating issues and problems in Pangani River basin.

1.2 RESEARCH OBJECTIVES

The goal of this research is to assess water use, potential opportunities and trade-offs in water allocation that can lead to increased water productivity and water use efficiency in a heterogeneous, highly utilized but data scarce African river basin. In the end, the study provides tools and information that enable policy makers, planners and stakeholders make well-informed decisions in integrated water resources planning and management to enhance socio-economic development and environmental sustainability in the river basin. New methodology approaches have been developed to capture the unique hydrological features of the landscape.

The specific objectives of the study are:

1. Develop at appropriate scales, the land use and land cover, and the spatial and temporal variability of evaporation and transpiration of the Upper Pangani River Basin. This provides the boundary conditions for water balance and water productivity analyses.

2. Develop a hydrological model for the Upper Pangani River Basin (Kikuletwa and Ruvu Catchments) that accounts for the distribution of green and blue water in time and space.

3. Develop spatially explicit water productivity maps for agricultural and natural landscapes in the Upper Pangani River Basin. The water productivity is presented using both biophysical (biomass and yield) and economic indices.

4. Develop an integrated hydro-economic model (IHEM) for green-blue water uses the entire Pangani River Basin. The IHEM is used to evaluate optimal policies and basin strategies for increased water productivity and environmental sustainability against the current institutional policies, priorities or preferences of the key stakeholders including the environment.

1.3 STRUCTURE OF THE THESIS

The thesis consists of eight chapters that can be categorized into four parts (Fig. 1.2).

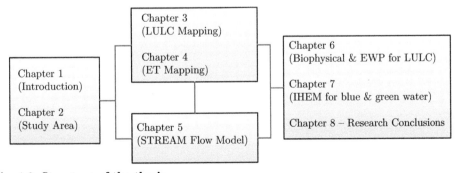

Fig. 1.2: Structure of the thesis.

Chapters 1 and 2 are introductory chapters. Chapter 1 provides the research setting, problem description and the research objectives. Chapter 2 provides an overview of the study area, the Pangani River Basin. Chapters 3 and 4 provide the boundary conditions for the study. Chapter 3 (Kiptala et al., 2013a) provides a detailed land use and land cover map for the study area. Chapter 4 (Kiptala et al., 2013b) presents the evaporation and transpiration fluxes and the water balance for the river basin. Chapter 5 presents a new hydrological model (STREAM) developed to account for green and blue water use in the Upper Pangani River Basin (Kiptala et al., 2014). Chapter 6 (Kiptala et al., 2016a) provides the biophysical and economic water productivity for agricultural and natural landscapes for the Upper Pangani River Basin. Chapter 7 (Kiptala et al., 2016b) integrates the green and blue water use and presents optimized options for improved water productivity and water value for the entire Pangani Basin. Chapter 8 presents a summary drawn from all the chapters of the thesis. The last chapter also highlights the study conclusions and limitations for further research and the contributions to science.

Chapter 2

STUDY AREA

The chapter presents an overview of the entire Pangani River Basin and the upper catchments that form the Upper Pangani River Basin.

2.1 LOCATION

The Pangani River Basin is a trans-boundary river basin, with major part in Tanzania and a small part in Kenya (Fig. 2.1). It is located between latitude 3 – 6° S and longitude 36 – 39° E in Eastern Africa. The river basin is made up of five main catchments: the Kikuletwa, Ruve, Mkomazi, Luengera and the Pangani mainstream. In total, Pangani River Basin has a drainage area of 43,000 km^2.

The Upper Pangani River Basin (13,400 km^2) covers approximately 30% of the total area of the Pangani River Basin. The Upper Pangani River Basin is the main headwater of the entire river basin and derives its water resources from Mt. Meru (4,565m) and Mt. Kilimanjaro (5,880m) catchments. These catchments are characterized by perennial springs which are fed from the mountains, then join at NyM reservoir. Irrigation development consumes most of the water resources in the sub-basin, up to 64% of the total blue water (World Bank, 2006). NyM reservoir (100 km^2), Lake Jipe (25 km^2), Lake Chala (5 km^2) and the expansive national parks (Tsavo West, Amboseli, Arusha and Kilimanjaro) are located on Upper Pangani River Basin.

The Lower Pangani River Basin comprises of mainly semi-arid plateau and some localized flow systems originating from the Pare and Usambara mountains. The river systems forms the Mkomazi and Luengera tributaries that join the Pangani river system then flows to the Pangani estuary, a total distance of 500 km. The Lower Pangani River Basin has three operational hydro-electric power (HEP) stations: NyM, Hale and the New Pangani Falls stations. These provide up to 91.5 MW or 17% of Tanzania's hydropower production which is about 11% of Tanzania's electricity supply. The river flow for hydropower production is regulated at the NyM reservoir, with a storage of 1.1×10^9 m^3. A large wetland, Kirua swamp, is also located in the lower basin and relies on the water supply from the Upper Pangani River Basin. The size of the wetland has reduced since the construction of the NyM reservoir.

2.2 CLIMATE

The high altitude slopes around the mountain ranges have an Afro-Alpine climate and receive nearly 2,500 mm yr^{-1} of rainfall. The lower parts have a sub-humid to semi-arid climate and the rainfall varies between 300 to 800 mm yr^{-1}. The rainfall has a bimodal pattern where long rains are experienced in the months of March to May (*Masika* season) and the short rains in November to December (*Vuli* season).

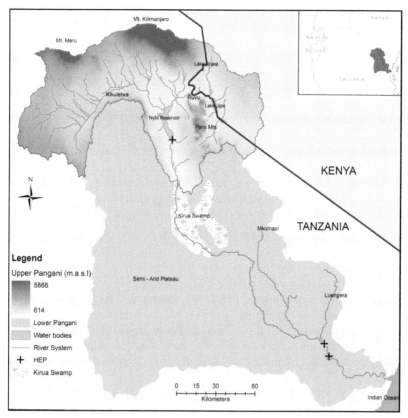

Fig.2.1: Location and overview of Pangani River Basin and the Upper Pangani River Basin.

2.3 SOCIO-ECONOMIC ACTIVITIES

Agricultural activities are predominant in the upper catchments while the lower catchments have limited but high potential for agricultural development, constrained by water scarcity. Livestock is dominant especially with the *Maasai* community in the dry plains in the lower catchments. Water resources are also utilized for hydropower, irrigation but also to sustain environmental resources such as wetlands and the estuary in the lower basin.

Chapter 3

LAND USE AND LAND COVER CLASSIFICATION[1]

In arid and semi-arid areas, evaporation fluxes are the largest component of the hydrological cycle, with runoff coefficient rarely exceeding 10%. These fluxes are a function of land use and land management and as such an essential component for integrated water resources management. Spatially distributed land use and land cover (LULC) maps distinguishing not only natural land cover but also management practices such as irrigation are therefore essential for comprehensive water management analysis in a river basin. Through remote sensing, LULC can be classified using its unique phenological variability observed over time. For this purpose, sixteen LULC types have been classified in the Upper Pangani River Basin (the headwaters of the Pangani River Basin in Tanzania) using MODIS vegetation satellite data. Ninety-four images based on 8 day temporal and 250 m spatial resolutions were analyzed for the hydrological years 2009 and 2010. Unsupervised and supervised clustering techniques were utilized to identify various LULC types with aid of ground information on crop calendar and the land features of the river basin. Ground truthing data were obtained during two rainfall seasons to assess the classification accuracy. The results showed an overall classification accuracy of 85%, with the producer's accuracy of 83% and user's accuracy of 86% for confidence level of 98% in the analysis. The overall Kappa coefficient of 0.85 also showed good agreement between the LULC and the ground data. The land suitability classification based on FAO-SYS framework for the various LULC types were also consistent with the derived classification results. The existing local database on total smallholder irrigation development and sugarcane cultivation (large scale irrigation) showed a 74% and 95% variation respectively to the LULC classification and showed fairly good geographical distribution. The LULC information provides an essential boundary condition for establishing the water use and management of green and blue water resources in the water stress Pangani River Basin.

---

[1] This chapter is based on: Kiptala, J. K., Mohamed, Y., Mul, M., Cheema, M. J. M., and Van der Zaag, P., 2013a. Land use and land cover classification using phenological variability from MODIS vegetation in the Upper Pangani River Basin, Eastern Africa. *Physics and Chemistry of the Earth*, 66, 112-122.

3.1 INTRODUCTION

Information on Land Use and Land Cover (LULC) is fundamental to water resources management. This information is used for the estimation of root zone depth, interception capacity and hydrotope delineation (Winsemius, 2009) and for computing evapotranspiration (ET) in a river basin (Cheema and Bastiaanssen, 2010). LULC influences the partitioning of rainfall into *green* (moisture in the soil) and *blue water* flows (water in rivers, lakes, dams, and groundwater). *Green water* flows is a subject of much interest in tropics and arid regions where it dominates the hydrological cycle. Management of *green water* flows requires explicit integration of land issues with water issues. However, this has been inadequate due to complexities in the estimation of water use of land based activities (Jewitt, 2006).

Several global and regional land cover maps have been developed using satellite information. For example the Food and Agricultural Organization of the United Nations (FAO) and the International Food Policy Research Institute (IFPRI) LULC maps developed in 1993 from 1 km Advance Very High Resolution Radiometer (AVHRR) and the Global Land Cover (GLC2000) developed in 2000 using 1 km Satellite Pour I' Observation de la Terra (SPOT) vegetation data. These global databases of low spatial resolutions (1 to 10 km) were produced primarily for global applications (Giri and Jenkins, 2005). As reported by the International Society for Photogrammetry and Remote Sensing (ISPRS, 2011), these databases lack adequate details at national or river basin scales and are of inadequate quality. Furthermore, such global databases cannot distinguish adequately specific crops and only detects dominate land covers leading to a large percentage of mixed classes with natural vegetation (Portman et al., 2010). As such, they cannot be used independently in considerably high heterogeneous catchments. Moreover, most information contained in these databases is also relatively old, most being developed more than 10 years ago.

Advances in remote sensing technology and geospatial data processing applications enable classification and updating of LULC maps with adequate accuracy at various scales (Cheema and Bastiaanssen, 2010; de Bie et al., 2011; Nguyen et al., 2012). Moderate-resolution Imaging Spectroradiometer (MODIS) vegetation images have been found to have better capabilities for land use classification with higher accuracy at a river basin scale (Giri and Jenkins, 2005; Fisher and Mustard, 2007). Presently, MODIS (Terra and Aqua) vegetation images are provided every 16 days at 250 m spatial resolution and can therefore provide 8 days time step for LULC analysis. The moderate resolution (250 m spatial and 8 day temporal) is reasonable good enough to support agricultural water management in the river basin.

Using remote sensing data only has also been found to produce results of lower accuracies. It has been found necessary to refine and improve the capabilities of satellite imagery with secondary information, such as cropping calendars (Zhang et al., 2008; de Bie et al., 2011; Klein et al., 2012). Recent studies using MODIS 250-m and secondary information for LULC classification at river basin scale obtained classification accuracies between 76 - 90% (Knight et al., 2006; Wardlow and Egbert, 2008; Zhang et al., 2008; Clark et al., 2012; Klein et al., 2012). Zhang et al. (2008) further did a comparative study between MODIS and Landsat Thematic Mapper data at a river basin scale and confirmed that MODIS datasets provided better classification accuracy. These studies were done in the USA and China with a comparable temperate humid climate conditions.

This chapter aims at deriving detailed and up-to-date LULC in a considerably heterogeneous and data scarce landscape in Eastern Africa. The study used 250 m MODIS vegetation images and secondary information on the growing pattern of crops and ground observations of dominant land features.

3.2   Materials and methods

3.2.1   Crop calendar

Cropping calendar provides key information for refining land use classification for managed agricultural practices. The crop calendar for Upper Pangani River Basin has been developed with the aid of local information from the Irrigation Department, Ministry of Water and Irrigation, Tanzania. Other general information considered were the general crop calendar patterns provided by FAO (FAO, 2011) and United States Department of Agriculture, Foreign Agricultural Services (USDA, 2011) for different climate conditions or countries. Irrigated crops, fruits and vegetables were cultivated throughout the year especially in the upper catchments. The cropping calendar for sugarcane grown in large scale plantation ensures that approximately 60% of the crops felled in the development stage during the *Masika season*. This was corroborated with field data that indicated the influence of the limited water resources during the dry season. The crop calendar provides for reduced operating cost (no pumping as the water levels in the river is high during the wet season) and also ensures the harvesting of sugarcane during the dry season.

Table 3.1: Crop calendar of Upper Pangani River Basin.

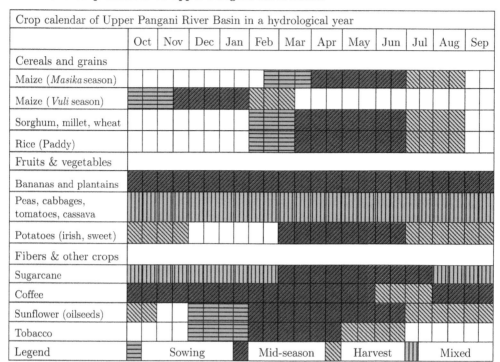

Crop calendar of Upper Pangani River Basin in a hydrological year												
	Oct	Nov	Dec	Jan	Feb	Mar	Apr	May	Jun	Jul	Aug	Sep
Cereals and grains												
Maize (*Masika* season)												
Maize (*Vuli* season)												
Sorghum, millet, wheat												
Rice (Paddy)												
Fruits & vegetables												
Bananas and plantains												
Peas, cabbages, tomatoes, cassava												
Potatoes (irish, sweet)												
Fibers & other crops												
Sugarcane												
Coffee												
Sunflower (oilseeds)												
Tobacco												
Legend		Sowing			Mid-season			Harvest			Mixed	

3.2.2   Pre-processing of the MODIS datasets

MODIS is an extensive program using sensors on two satellites (Terra and Aqua) to provide global observations of the Earth's land in the visible and infrared regions of the spectrum. Terra satellite was launched in 1999 while Aqua was launched in 2002. The MODIS data is available in different versions, and the latest version 5 (V005) available from 2008 from USGS database have been validated (USGS, 2012). The images were obtained freely from the Land Processes Distributed Active Archive Center (LPDAAC) of the National Aeronautics Space Administration (NASA), [https://reverb.echo.nasa.gov/reverb].

The MODIS vegetation products were converted the Normalized Difference Vegetation Index (NDVI) by dividing with 10,000. To have continuous satellite data, cloud pixels in the images have be cleaned using advanced interpolation techniques in ERDAS imagine software (ERDAS, 2010). For each image with cloud pixels, an area of interest (AOI) was created over the clouded area (only the section of the image with cloud damage). If the AOI has not been completely damaged by the clouds, the pixels that have correct spectral values were randomly picked and interpolated over the AOI. If the AOI has been fully cloud damaged, the histogram matching option was used to match data with the adjacent scene (assumed to have similar spectral characteristics) or same scene from the next or previously available image. This was critical for Upper Pangani River Basin where cloud damaged pixels occurs mainly in the mountainous areas.

To minimize uncertainty from the interpolation procedure, a longer timeseries of data were analyzed covering both a relatively dry year (2009) and an average year (2010). This was also aimed at achieving better classification of the managed land use practices. In total 94 MODIS Terra/Aqua NDVI images were analyzed covering two hydrological years, Oct 2008 to Sep 2010 over the Upper Pangani River Basin.

3.2.3 Unsupervised and supervised classification

Remote sensing technology using satellite imagery can be used to observe and monitor vegetation density. The spectral reflectance depends on vegetation foliage which varies for particular crop or vegetation, and the crop growth stages over time. Healthy vegetation (green leaves) absorbs most of radiation in the visible and reflects very well in the near infrared part of the spectrum. The magnitude of NDVI (which represent the greenness of vegetation) is therefore related to the level of photosynthetic activity of the vegetation cover.

$$NDVI = \frac{(NIR - VIS)}{(NIR + VIS)} \tag{3.1}$$

where VIS and NIR are the spectral reflectance measurements in the visible (red) and near-infrared regions, respectively.

Using unique seasonal cycles of the vegetation types obtained from NDVI time profile (growth phenology), different LULC types can be identified using the unsupervised and supervised classification technique. The unsupervised classification has been used initially to create a thematic raster layer using their spectral similarities (from the statistical patterns in the data), while defining the appropriate clustering sample. ISODATA (Iterative Self Organizing Data Analysis Technique) and the k-mean are the commonly used unsupervised classification algorithms in remote sensing. ISODATA is based on Euclidean distance, in which spectral distances between candidates pixels are compared to each cluster mean (Cheema and Bastiaanssen, 2010). The ISODATA algorithm has some further refinements by splitting and merging of clusters (Jensen, 1996). New cluster centers are computed by averaging the locations of all the pixels assigned to that cluster (Campbell, 2002). The entire process is repeated and each candidate pixel is compared to the new cluster means and assigned to the closest cluster mean. The ISODATA algorithm is also successful at finding the spectral clusters that are inherent in the data if enough iteration is allowed or a certain convergence threshold is achieved. A convergence threshold (confidence level) of 98% was adopted for this study.

The classification was later refined with expert judgement of crop calendar and land features using the supervised classification. In the supervised classification both the parallelepiped and minimum distance are used in evaluating signature files and refining the classification (ERDAS, 2007). The knowledge of the cropping pattern assists in defining specific NDVI temporal profiles and thus the signature files for different LULC types. This methodology is computationally intensive and ERDAS Imagine 9.2 software has been used in the study.

A comprehensive classification accuracy assessment has been adopted since there is no up-to-date land use map of fair resolution in the study area. It include: ground truthing, validation with local datasets for individual land use types and land suitability assessment. The classification accuracy has been evaluated and related to acceptable levels based on literature that have been deemed sufficient for water management analysis at a river basin scale.

3.2.4　Calibration and Validation

Ground truthing

To determine the quality of the LULC map generated, an error matrix approach was adopted, which uses the independent classification (from the LULC map) and ground or reference data (from ground truthing). A ground truthing survey was carried out from November - December 2010 to capture the mid *Vuli* season and from May - June 2011 to capture the mid *Masika* season. Sample size is key consideration in assessing the accuracy of the LULC map. Because of the large number of pixels in the LULC map (approximately 200,000 pixels), statistical methods of determining required sample size would lead to large no. of samples which is not practically feasible. A balance therefore between what is statistically sound and practicable attainable must therefore be achieved (Congalton, 1991). A general norm in assessing accuracy of remote sensed images is to collect a minimum of 3 samples for each land use category and the number adjusted upwards based on the relative importance of the land use within the objectives of the mapping. A minimum of three random observation points were therefore taken at a point where the class observed is approximately 70% of the dominant LULC type. In total, 253 samples were randomly sampled using Global Positioning System that covered 14 land use classifications except for the wetlands and swamps (geographically distinct & inaccessible) and the urban areas that was classified using different methodology (Table 3.3). It was therefore not possible to do random stratified sampling for all LULC types. According to Pouliet et al. (2012), the random sampling can still achieve high overall accuracy because the samples ensures that the most frequently classes are well characterized.

A larger sample size was therefore achieved for dominant land use types under mixed crops with the irrigated mixed crops, grasslands with scattered croplands and rainfed maize and beans having 44, 45, and maximum of 67 samples respectively. It is noteworthy that the ground truthing survey for the *Masika* season is not contemporaneous with the date of the images used, however no major changes on the land use practice is expected or was observed within the time interval (less than 1 year). Fig. 3.1 shows the ground truthing positions and some of the salient features of the Upper Pangani River Basin.

The sample data was summarized in an error matrix for the 253 observation points that was subjected to accuracy assessment based on two procedures (i) Overall classification accuracy; and (ii) Kappa statistic.

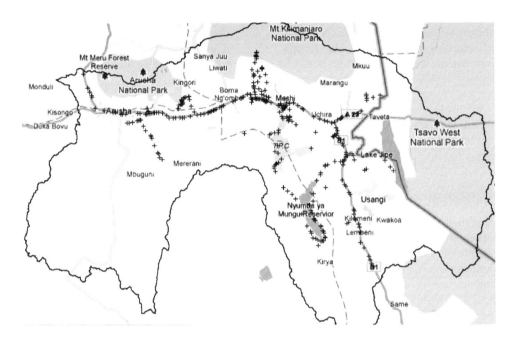

Fig. 3.1: Ground truthing positions and some salient features of Upper Pangani River Basin.

Overall classification accuracy

The overall classification accuracy (AC_o) was derived by dividing the total number of correctly classified landuse classes by the total number of reference data.

$$AC_O = SQ_{cc} / SQ_{tc} \qquad (3.2)$$

where SQ_{cc} is the total number of sampling classes classified correctly, and SQ_{tc} is the total number of reference sampling classes.

The individual accuracy of the LULC types can also be estimated using the producer's accuracy and user's accuracy. The producer's accuracy has been computed by dividing the number of samples in an individual class identified corrected by the respective reference totals while the user's accuracy has been computed by dividing the number of samples in an individual class identified corrected with the classified totals (Lillesand and Keifer, 1994; Townshend, 1981). The overall classification accuracy is the overall mean of the producer's and user's accuracy.

Kappa statistic

A better statistical index to determine classification accuracy is the Kappa statistic, which expresses the agreement between two categorical datasets corrected for the expected agreement (Van Vliet et al., 2011). The Kappa statistic incorporates the off-diagonal elements of the error matrices and eliminates class assignment by chance (Congalton et al., 1983). The Kappa (K) coefficient will equal 1 if there is perfect agreement, whereas 0 is what would be expected by total chance alone.

$$K = \frac{d - q}{N - q} \qquad\qquad (3.3)$$

where d is the overall value for percentage correct, q is the estimate of the chance agreement to the observed percentage correct (calculated from the number of cases expected in diagonal cells by chance) and N is the total of number of cases.

Validation with local datasets

Regional LULC maps available, for example the Global Land Cover (GLC2000) or other sources such as available at the FAO website were developed using coarse resolution imagery (1 km - 10 km). An existing land cover map (with a resolution 1 km) for the entire Pangani River Basin had a similar shortcoming (IUCN, 2003). The information contained in these maps is therefore too coarse to be compared with the LULC developed in this study. However, some studies on individual land use types were available for validation. A recent irrigation survey for the Upper Pangani River Basin was done by the irrigation department, Ministry of Water and Irrigation, to assess the irrigation development in the basin (MOWI, 2009). The survey covered both smallholders and large scale irrigation and was undertaken for the period November 2008 - May 2009. Other information available were the surface areas of the water bodies viz; the NyM reservoir, Lake Jipe and Lake Chala in the Upper Pangani (IUCN, 2003; PBWO/IUCN, 2008; IUCN, 2009).

Land suitability to LULC types

Vegetation can also be evaluated by considering landscape features that are ideal for crop developments. Physical and climatic (seasonal) factors influence the vegetation growth in a similar way as observed in the NDVI values for various LULC types. These factors when considered simultaneous will provide land suitability indicators for agricultural development (as well as natural vegetation). FAO-SYS (FAO, 1976; 1983; 2007) provided a framework for evaluating land suitability for agricultural development based on vegetation indices. The FAO-SYS system uses climate, topography and soil to evaluate and indicate degree of land suitability to certain crops or vegetation. The FAO-SYS framework has recently been improved to use spatial data for better evaluation of larger land masses e.g. at river basin scale. Such recent application includes the Agricultural Land Suitability Evaluator (ALSE) developed for tropical and subtropical climate (Elsheikh et al., 2013).

In a simplified way, the land suitability index has been used to evaluate the LULC types using limited but key parameters derived from FAO-SYS for the Upper Pangani River Basin. The key suitability parameters used include: climate (precipitation), topography (slope and elevation) and soil (soil depth). An ideal parameter range has been chosen for each parameter based on FAO-SYS and the local environmental conditions. The ideal range has been assign suitability factor of 1. The suitability degree decreases proportionally to the actual parameter values. The ideal parameters used for this study to represent natural vegetation were based on the moderately good conditions for growth of rainfed maize, coffee, mango and bananas in tropical climate (FAO, 1983; Elsheikh et al., 2013). They include: climate (annual precipitation of > 1500 mm yr^{-1}), topography (elevation < 2000 m.a.s.l and slope < 15%) and soils (soil depth > 150 mm).

The precipitation data was obtained from 43 rainfall stations located in the Upper Pangani River Basin. The point measurement data were interpolated using the inverse distance method to generate a precipitation map for Upper Pangani River Basin. The elevation and slope were extracted from a 90 m resolution digital elevation map (DEM) obtained from the Shuttle Radar Topographic Mission (SRTM) database (Jarvis et al., 2008). The soil map was obtained from the harmonized world soil database which relied on soil and terrain (SOTER) regional maps for Northern Africa and Southern Africa (FAO/IIASA/ISRIC/ISS-CAS/JRC, 2012). The suitability assessment was done on a similar scale of 250 m using the spatial data.

The overall suitability was obtained by multiplying the suitability factor for each parameter.

$$S = S(c) \times S(t) \times S(s) \tag{3.4}$$

Where S is the overall suitability, $S(c)$ is suitability factor for climate (precipitation), $S(t)$ suitability factor for topography (slope and elevation) and $S(s)$ suitability factor for soil (soil depth).

The overall suitability of each LULC type has been evaluated on 5 suitability classes identify in FAO-SYS. These classes are: $S_1 = 0.85$ (suitable), $S_2 = 0.60$ (moderately suitable), $S_3 = 0.45$ (marginally suitable), $N_1 < 0.45$ (not suitable), and $N_2 < 0.45$ (Not suitable for physical reasons). The computations were scripted in PCRaster modelling environment. A GIS based software with a rich set of model building blocks and analytical functions for manipulating Raster GIS maps (Karssenberg et al., 2001).

3.3   RESULTS AND DISCUSSION

3.3.1   Land surface phenology

The clustering of various LULC types was initially undertaken using unsupervised classification where 40 clusters (estimated to be twice the LULC types from field observations) where generated. This preliminary classification provided the statistical basis for further refinement using the supervised classification. The croplands were identified using the NDVI temporal profiles and an expert judgment of the cropping calendar. Other natural land cover classes were also identified based on ground information on the land features, location and expert judgment of their vegetative or NDVI profile patterns. The supervised classification resulted in the reduction of the initial 40 clusters of land use types to a final classification of 15 classes.

The mean NDVI time profiles for seven agricultural classes are shown in Fig. 3.2. The NDVI values were derived from the zonal mean of the NDVI for individual classes within the domain (Upper Pangani) for all the 94 time steps. A three period moving average filter was used to smoothen the profile by introducing a time lag using one period before and one period after the time of analysis (time step) as described by Reed et al. (1994). The smoothing of the profile was also necessary to minimize the effects of interpolated cloud damaged pixels in some images.

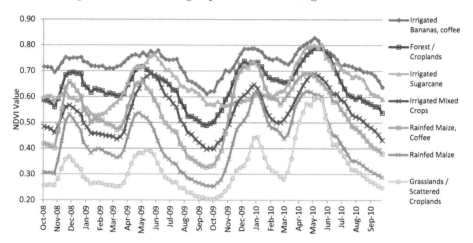

Fig. 3.2: Mean NDVI curves for Irrigated and Rainfed Croplands in Upper Pangani River Basin for the hydrological years 2009 and 2010.

The peaks on the agricultural classes were quite distinctive of the cropping seasons, the *Masika* and *Vuli* seasons. The rainfed croplands tend to follow the cropping seasons with relatively higher peaks during the long *Masika* seasons and relatively lower peaks during the short *Vuli* seasons. The irrigated areas and croplands on the uplands (mixed with forest) have enhanced foliage cover compared to rainfed croplands and has been revealed by the NDVI time profile. Since supplementary irrigation is generally practiced, the vegetative pattern also follows the rainfall pattern in the river basin, though less distinct compared to rainfed crops.

Fig. 3.3 shows the mean NDVI time curves for eight natural land cover classes. These classes include forests, wetlands and swamps, water bodies, and pastures and the savannas. Wetlands and swamps, dense forest and afro-alpine forest have high NDVI values (0.6 - 0.8) because of their high vegetation cover. Afro-Alpine forest vegetation has been suppressed by the lower temperatures at the higher elevations of Mt. Kilimanjaro and Mt. Meru. However, during the 'summer' months of January and February, when the temperatures were higher, the vegetation foliage was enhanced as shown in Fig. 3.3. Afro-alpine forest forms a transition from the dense forest to the bareland/ice in the mountains peaks.

The wetlands and swamps also receive blue water from river systems and ground water flows to maintain the high vegetation density. The seasonal variability of flow also influences the water availability for the wetlands. This results in lower NDVI values during the dry seasons (NDVI of 0.6 from 0.8 in wet seasons) for the wetlands and swamps.

The NDVI values for the other natural land covers viz; shrublands and or thicket, bushlands and sparse vegetation, were significantly influenced by the rainfall pattern in the river basin.

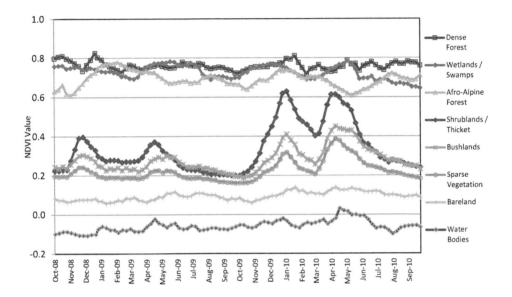

Fig. 3.3: Mean NDVI curves for natural land cover in Upper Pangani River Basin for the hydrological years 2009 and 2010.

Fig. 3.4 shows representative rainfall pattern of three stations distributed in the Upper Pangani River Basin. Arusha (airport) and Moshi (airport) rainfall stations are located at the foot of Mt. Meru and Mt. Kilimanjaro catchments. Same (meteorological) rainfall station is located in the lower parts of the catchment (see Fig. 3.1). The hydrological year 2009 was dry with relatively low rainfall especially on the lower catchments (Same Station) compared to the hydrological average year 2010. This has influenced the vegetative growth or health of the grasslands, shrublands and bushlands as shown in Fig. 3.3 and 3.4.

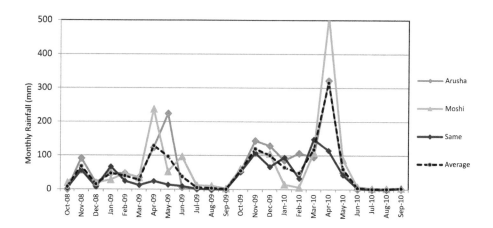

Fig. 3.4: Monthly Rainfall distribution for 3 stations in the Upper Pangani River Basin.

The other landuse feature considered in the classification was the urban or built-up areas. Though MODIS data has a better ability to discriminate urban area (Giri and Jenkins, 2005) they are always classified as mixed pixels (barelands or sparse vegetation) depending on their vegetative patterns of the built-up areas. Furthermore, the moderate resolution of MODIS data (250 m) might be too course to adequately discriminate urban or built-up areas from the adjacent land use types. Consequently, other methodologies have been derived to deal with this problem including using higher resolution images (Zhang et al., 2008). In this chapter, Google maps for the two main urban areas in the basin, Arusha and Moshi, were used to map out the class and mask over the LULC map.

The areal extents of the LULC classes are summarized in Table 3.2 and shown in Fig. 3.5.

Shrublands or thicket with coverage of 26.3% of the total area was the dominant land use type. The dominance has been influenced by the expansive Tsavo West National Park (classified as shrublands) on the lower eastern part of the river basin. Rainfed maize (22.1%), grasslands with scattered croplands (11.4%) and bushlands (8.6%) also constitute large coverage. In irrigated agriculture, bananas and coffee dominates the upper slopes of the catchments (4.6%) while mixed crops (maize, paddy, bananas, and vegetables) dominate the mid slopes of the catchments (4.5%). The least LULC include irrigated sugarcane, water bodies, bareland, wetlands and swamps all at 0.7% and urban (0.1%) of the total catchment area.

The classification also captured some of the main salient features of the river basin (Fig. 3.2). Mt. Kilimanjaro and Arusha National Parks were classified as natural land cover consisting of afro-alpine forest, sparse vegetation, and barelands (including the ice caps) at the higher elevations. The sugarcane plantation, Tanzania Plantation Company (TPC) has also been classified correctly as Irrigated Sugarcane.

Table 3.2: LULC classes and their areal distribution in Upper Pangani River Basin.

No.	Land Use	Area (km²)	Area (%)	
1	Water bodies	100	0.7%	
2	Bareland	100	0.7%	
3	Sparse vegetation	445	3.3%	
4	Bushlands	1,152	8.6%	
5	Grasslands and scattered croplands	1,517	11.4%	
6	Shrubland and or thicket	3,509	26.3%	
7	Rainfed maize	2,942	22.1%	
8	Afro-Alpine forest	257	1.9%	
9	Irrigation mixed crops	598	4.5%	
10	Rainfed coffee, banana	723	5.4%	
11	Irrigation, sugarcane	89	0.7%	
12	Forest, croplands	556	4.2%	
13	Irrigation; banana, coffee	607	4.6%	
14	Dense forest	637	4.8%	
15	Wetlands and swamps	98	0.7%	
16	Urban and built-up	8	0.1%	
	Total	13,337		

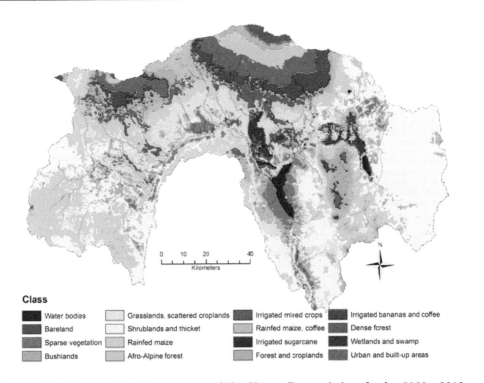

Fig. 3.5: Land use and land cover map of the Upper Pangani river basin, 2009 - 2010.

3.3.2 Ground truthing

An error matrix report for the ground truthing has been summarized and presented in Table 3.3. The results show an overall classification accuracy of 85%, with the producers' accuracy of 83% and users' accuracy of 86% for a confidence level of 98% (conveyance threshold) of analysis. The accuracy levels for individual classes showed relatively good accuracies of more than 70% expect for the barelands (60%). The low producer's accuracy can be associated with the existence of more than one LULC in the pixels used in the ground truthing. For that reason, the user's accuracy for barelands was higher at 100%. A similar condition was also observed for the classification accuracies for rainfed maize and irrigated sugarcane.

The selection of the reference class within the pixel (which was assigned by chance) might not have been identified correctly for these LULC types. The uncertainty has also been corrected by the Kappa statistic (K). K increased the accuracy for bareland from 0.6 to 1.0. Similarly, the sugarcane cultivation had its accuracy corrected from 0.83 to 1.0. Typically, K values greater than 0.8 represent strong agreement, K values between 0.4 - 0.8 represent moderate agreement and K values below 0.4 are indicative of poor agreement between the remotely sensed classification and the reference data (Landis and Koch, 1977; Congalton and Green, 1999). Most land use studies using various types of satellite imagery have attained overall Kappa coefficients of between 0.64 and 0.89 (Van Vliet et al., 2011). Examples from previous studies using 250 m MODIS include: Jonathan et al. (2006) with a overall Kappa of 0.72 in a study in a river basin in Brazil; Zhang et al. (2008) attained an overall Kappa of 0.66 in the North China Plains, Tingting and Chuang (2010) achieved a Kappa of 0.80 in Chao Phraya River basin in Thailand and a study in the Bolivian seasonal tropics by Redo and Millington (2011) attained a Kappa of 0.87. The overall Kappa coefficient achieved for the study was 0.85, which represents a good agreement between the LULC and the ground information.

In the overall classifications accuracy, Bastiaanssen (1998) reported that crops can be classified with an accuracy of 86%. Thunnisen and Noordman (1997) suggested at a regional scale a minimum overall classification accuracy of 70%. Ozdogan and Gutman (2008) attained an overall accuracy of between 79% - 87% for various river basin in continental US using 500 m MODIS, Wardlow and Egbert (2008) attained 84% accuracy using 250 m MODIS in the US Great plains. Cheema and Bastiaanssen (2010) attained 77% overall accuracy using 1 km SPOT in the Indus basin while Nguyen et al. (2012) attained higher accuracy of 94% mapping rice fields using hyper-temporal SPOT images in the Mekong delta. This study's overall classification accuracy of 85% was within acceptable levels ranges suggested and attained by the previous studies.

Table 3.3: Error matrix for LULC in the Upper Pangani River Basin.

S/No.	Class Name	Reference Totals	Classified Totals	Number Correct	Producers Accuracy	Users Accuracy	Kappa (K)
1	Water bodies	3	4	3	100%	75%	0.75
2	Bareland	5	3	3	60%	100%	1.00
3	Sparse vegetation	23	22	20	87%	91%	0.90
4	Bushlands	12	12	9	75%	75%	0.74
5	Grasslands/scatt. crops	45	47	37	82%	79%	0.74
6	Shrublands and or thicket	6	7	5	83%	71%	0.71
7	Rainfed maize	67	65	53	79%	82%	0.75
8	Afro-Alpine forest	4	4	4	100%	100%	1.00
9	Irrigated mixed crops	44	49	38	86%	78%	0.73
10	Rainfed coffee, bananas	7	5	5	71%	100%	1.00
11	Irrigated sugarcane	6	5	5	83%	100%	1.00
12	Forest and croplands	5	6	4	80%	67%	0.66
13	Irrigated bananas, coffee	13	10	10	77%	100%	1.00
14	Dense forest	13	14	13	100%	93%	0.92
	Totals	253	253	209	83%	86%	0.85

Overall classification accuracy = 85%; Overall kappa statistics = 0.85

3.3.3 Validation with local datasets

Table 3.4 shows the comparison of LULC map with the available datasets for specific land use types in Upper Pangani River Basin. Total irrigation in the LULC resulted in a 74% agreement to the local datasets at the Ministry of Water and Irrigation (MOWI). The moderate agreement may be attributed to the different methodology used that is subject to varying uncertainties and accuracy in the classifications. The notable difference was how informal supplementary irrigation has been defined and assessed. In the classification, irrigated area including informal supplementary irrigation has been identified using vegetation growth assessed using satellite imagery and limited ground data. The main uncertainty for this classification arises from the moderate resolution of the images (6.25 ha). These pixels might not be contemporaneous with some smallholder farms which can lead to mixed classes. However, for large scale classification e.g. at river basin scale, these errors are likely to cancel out at the end.

Table 3.4: Comparison of some land use types with other sources of datasets.

S/No.	Classification	Present study, (ha)	Other sources (ha)	% agreement	Source
1	Total irrigated area	129,406	95,823	74	MOWI (2009)
2	Irrigated sugarcane	8,919	8,480	95	MOWI (2009)
3	Water bodies	10,525	7,555 – 18,800	72 - 179	IUCN (2003), PBWO/IUCN (2008)

Conversely, the local datasets using field surveys tend to underestimate the irrigated areas especially for smallholder informal supplementary irrigation. Supplementary irrigation especially smallholder is always spontaneous, unregulated and unorganized. It is therefore difficult to assess adequately during field survey. Unregulated irrigation using cheap suction pumps were also observed along the river banks. The level of uncertainty is therefore higher with this methodology which may have resulted to the lower irrigation estimates.

The large irrigation project (TPC sugarcane) provided a higher degree of separability where a good agreement between the classification and the local datasets of 95% was observed. Water bodies provided an agreement of between 72% and 179% compared to various studies on the variation of the surface areas of the NyM reservoir, Lake Jipe, and Lake Chala. The large variability can be attributed to large changes in the sizes of water bodies compared to the mean (nonlinearity) especially during wet and extreme dry seasons. Such observation was attributed to the NyM reservoir, a shallow dam that has large water level fluctuation (PBWO/IUCN, 2008).

3.3.4   Land suitability to LULC types

Fig. 3.6 shows the land suitability map for crop production in the Upper Pangani River Basin. Table 3.5 provides the overall suitability index for various land use types. Dense forest and irrigated bananas, coffee land use types have attained the highest suitability level (S_1). The natural shrublands (dominated by protected national parks), rainfed and irrigated croplands, and forested areas (afro-alpine and mixed with croplands) and urban also attained moderately high suitability level (S_2). These land use types have moderately high vegetation density and were located on the middle upper catchments of the river basin. The land suitability at the mountain peaks was significantly low due to the topographical parameters (elevation and slope) that were not favourable crop or plant development.

The land use types on the lower catchments: barelands, sparse vegetation, grasslands and also irrigated sugarcane attained marginally lower suitable level (S_3). Precipitation factor was the limiting parameter. Irrigated sugarcane score of 0.39 was attributed to the low precipitation factor (0.39). The areas have been considered to have high irrigation potential since precipitation is the only limiting factor for agricultural development (World Bank, 2006).

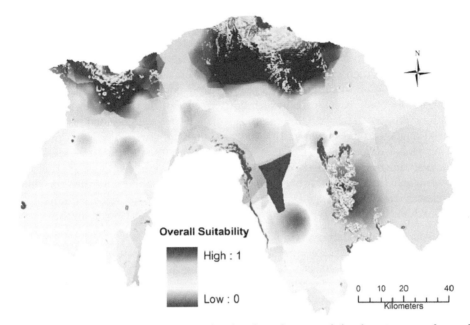

Fig. 3.6: Land suitability for crop production based on precipitation, topography and soils for Upper Pangani River Basin.

Table 3.5: Land suitability ratings based on precipitation, topography and soils for various land use types in Upper Pangani River Basin.

Land use type	Precipitation	Topography	Soil	Overall Suitability (S)	Suitability class
Water bodies	0.44	1.00	0.55	0.26	N_1
Bareland/ice	0.81	0.58	0.85	0.34	S_3
Sparse vegetation	0.44	0.97	0.90	0.37	S_3
Bushlands	0.49	0.94	0.97	0.42	S_3
Grasslands & few crops	0.46	1.00	0.99	0.45	S_3
Natural shrublands	0.53	1.00	0.99	0.53	S_2
Rainfed, maize and beans	0.52	0.99	0.99	0.51	S_2
Afro-alpine forest	0.99	0.70	1.00	0.69	S_2
Irrigated mixed crops	0.60	0.99	0.99	0.59	S_2
Rainfed coffee, banana	0.68	0.89	0.98	0.60	S_2
Irrigated sugarcane	0.39	1.00	1.00	0.39	S_3
Forest, croplands	0.76	0.87	0.99	0.65	S_2
Irrigated bananas, coffee	0.91	0.94	0.99	0.85	S_1
Dense forest	0.96	0.90	0.98	0.84	S_1
Wetlands & swamps	0.47	1.00	0.94	0.44	S_3
Urban and built-up	0.65	1.00	1.00	0.65	S_2

Rating: $S_1 = 0.85$ (suitable); $S_2 = 0.60$ (moderately suitable); $S_3 = 0.45$ (marginally suitable); $N_1 = 0.25$ (Not suitable); $N_2 = 0$ (Not suitable for physical reasons) (FAO, 1976; 1983; 2007).

3.4 CONCLUSION

This chapter has used recent advancement in remote sensing and satellite imagery to develop a spatially distributed and up-to-date LULC map. The LULC map developed at moderate resolution of 250 m using MODIS vegetation satellite data would provide essential information to support basin scale water resource management. The research combined seasonal phenological variation received from satellite imagery with secondary ground data such as cropping calendar to classify land use types that included managed land use practices that could not otherwise have been provided by existing global cover maps. The increased temporal resolution of the MODIS data and longer time series ensured actual timing of change event in the vegetation growth that were matched by the expert knowledge of the cropping calendar for different crop classes. However, change events for the urban areas were not discriminated from other land use types and were classified using high resolution Google maps.

The research attained an overall classification accuracy of 85%, with the producer's accuracy of 83% and user's accuracy of 86%. The Kappa coefficient of 0.85 is in good agreement range for land use classification. The Kappa also provided improved agreement for individual classes identified by chance in the error matrix. The overall classification for smallholder agricultural land showed a moderate agreement of 74% to the local statistics available in the river basin. The moderate agreement was attributed to the existence of informal supplementary irrigation and the uncertainties of the moderate scale (250 m) to fully identify agricultural plots of smaller sizes. The classification on sugarcane plantation, the only large scale irrigation in the river basin resulted in good agreement of 95% to actual field estimates. The land suitability classifications based on FAO-SYS framework provided the optimum suitability of the LULC to the production of major crops (e.g. maize, bananas, coffee and mango). The integration of the spatial geo-environmental conditions (climate, soils and topography) that characterize the land suitability and the availability water resources could give explicit indicators of the agricultural potential of the river basin.

The LULC classification provides an essential boundary condition for establishing the water use and management of *green* and *blue* water resources in the river basin. This is particularly crucial for a complex river system such as Pangani River Basin that has intensively managed landscapes (e.g. irrigation) in a bimodal tropical climate that is associated with high evaporative water use. This information would provide essential ingredients for water accounting where beneficial and non-beneficial water resources are evaluated for improved water productivity in the river basin.

Chapter 4

MAPPING EVAPOTRANSPIRATION USING MODIS AND

SEBAL[2]

Evapotranspiration (ET) accounts for a substantial amount of the water use in river basins particular in the tropics and arid regions. However, accurate estimation still remains a challenge especially in large spatially heterogeneous and data scarce areas including the Upper Pangani River Basin in Eastern Africa. Using multi-temporal Moderate-resolution Imaging Spectroradiometer (MODIS) and Surface Energy Balance Algorithm of Land (SEBAL) model, 138 images were analyzed at 250-m, 8-day scales to estimate actual ET for 16 land use types for the period 2008 to 2010. A good agreement was attained for the SEBAL results from various validations. For open water evaporation, the estimated ET for Nyumba ya Mungu (NyM) reservoir showed a good correlations ($R = 0.95$; $R^2 = 0.91$; Mean Absolute Error (MAE) and Root Means Square Error (RMSE) of less than 5%) to pan evaporation using an optimized pan coefficient of 0.81. An absolute relative error of 2% was also achieved from the mean annual water balance estimates of the reservoir. The estimated ET for various agricultural land uses indicated a consistent pattern with the seasonal variability of the crop coefficient (K_c) based on Penman-Monteith equation. In addition, ET estimates for the mountainous areas has been significantly suppressed at the higher elevations (above 2,300m.a.s.l.), which is consistent with the decrease in potential evaporation. The calculated surface outflow (Q_s) through a water balance analysis resulted in a bias of 12% to the observed discharge at the outlet of the river basin. The bias was within 13% uncertainty range at 95% confidence interval for Q_s. SEBAL ET estimates were also compared with global ET from MODIS 16 algorithm ($R = 0.74$; $R^2 = 0.32$; RMSE of 34% and MAE of 28%) and comparatively significant in variance at 95% confidence level. The inter-seasonal and intra-seasonal ET fluxes derived have shown the level of water use for various land use types under different climate conditions. The evaporative water use in the river basin accounted for 94% to the annual precipitation for the period of study. The results have a potential for use in hydrological analysis and water accounting.

[2] This chapter is based on: Kiptala, J. K., Mohamed, Y., Mul, M. L., and Van der Zaag, P., 2013b. Mapping evapotranspiration trends usingMODIS and SEBAL model in a data scarce and heterogeneous landscape in Eastern Africa, *Water Resour. Res.*, 49, 8495–8510, doi:10.1002/2013WR014240.

4.1 INTRODUCTION

Evaporation (E) and transpiration (T) (jointly termed as evapotranspiration (ET)) accounts for a substantial amount of the water use in river basins particular in semi-arid savannah regions. Because of the spatial heterogeneity and temporal variability in water availability in these regions, water managers responsible for planning and allocating water resources need to have a thorough understanding of the spatial and temporal rates of ET. This information helps to better understand evaporative depletion and to establish a link between land use, water allocation, and water use in a river basin (Bastiaanssen et al., 2005). River basins such as the Upper Pangani River Basin typically have many different land use and land cover (LULC) types which transmit water as ET. The LULC types have changed over time, due to socio-economic factors, impacting on the water flows and ecosystem services in the downstream catchments.

Rainfall is partitioned into *green* (moisture in the soil) and *blue water* flows (rivers, lakes, dams, groundwater) (Rockström et al., 2009). Small changes in ET and hence the *green water* can result in major impacts on downstream *blue water* flows. The management of *green water* flows requires explicit understanding of the biophysical characteristics of the LULC types and associated spatiotemporal variability of water use. However, the estimation of ET has been inadequate due to complexities of estimating the actual water use of land based activities including irrigated agriculture and the cultivation of crops during the rainy seasons that receive supplementary irrigation (Jewitt, 2006). In addition, conventional methods of estimation of ET (pan, lysimeter, Bowen ratio, eddy correlation or the aerodynamic techniques) require detailed meteorological data that may not be available at the desired spatial and temporal scales. In-situ measurements are constrained in generating areal estimates both in terms of cost and accuracy because of natural heterogeneity and the complexity of hydrological processes in river basins. Moreover, in-situ procedures are time consuming if observations are to be made repeatedly to assess the temporal variability of ET.

The remote sensing approach using models like TSEB (Norman et al., 1995), SEBAL (Bastiaanssen et al., 1998a; 1998b), S-SEBI (Roerink et al., 2000) and SEBS (Su, 2002) have shown great potential in estimating ET over large areas using limited meteorological data. ET links the water balance to the surface energy balance with the heterogeneity of the landscape being accounted by the remote sensed data. The recent advancements in the availability of satellite images of finer to medium resolutions (spatial and temporal) have further enhanced its application potential. Medium resolution satellite images, e.g. the Moderate-resolution Imaging Spectroradiometer (MODIS) vegetation products, have capability to derive physical parameters for surface energy balance models at catchment or river basin scale (Batra et al., 2006; McCabe and Wood, 2006; Zhang et al., 2008). They are also freely available from two sensors (Terra and Aqua) thus enhancing its temporal resolution.

SEBAL and the Simplified Surface Energy Balance Index (S-SEBI) make use of the spatial variability of the surface temperature and reflectance, and vegetation index observations (Mohamed et al., 2004; Romaguera et al., 2010). On the other hand, Surface Energy Balance System (SEBS) and Two-Source Energy Balance (T-SEB) are physically based models that use an excess resistance term that accounts for roughness lengths for heat and momentum that are different for canopy and soil surface (Van der Kwast et al., 2009). These models have been applied with indicative *ET* of acceptable accuracies in different river basins under different climatological conditions. The SEBAL model in particular, has been widely applied in the tropical climate and more importantly in data scarce river basins in Africa (Farah and Bastiaanssen, 2001; Timmermans et al., 2003; Mohamed et al., 2004; Kongo et al., 2011).

Table 4.1 presents SEBAL applications and the validation efforts in various landscapes similar to the Upper Pangani River Basin. A bias range of between 4 and 26%.

Table 4.1: Surface Energy Balance Algorithm for Land (SEBAL) applications and means of validation on various landscapes.

Source	Location	No. of images	Length (Time)	Image type and spatial resolution	Land use types	Elevation range (m.a.s.l)	Means of Validation	Bias range
Farah and Bastiaanssen (2001)	Kenya	10	1 month	NOAA-AVHRR 1 km	Savannah	1,900 - 3,200m	Bowen Ratio	16%
Bastiaanssen and Bandara (2001)	Sri Lanka	3	3 years	Landsat 30m	Irrigated croplands	200 - 600m	Water balance	4%
Timmermans et al. (2003)	Botswana	1	1 day	MODIS 1km	Savannah	1,000m	Scintillometer	14%
Hemakumara et al. (2003)	Sri Lanka	10	5 months	Landsat 30m	Irrigated rice, palm trees	100m	Scintillometer	17%
Mohamed et al. (2004)	Sudan	37	12 months	NOAA-AVHRR 1km	Wetlands	200 - 1,400m	Water balance	4%
Zwart and Bastiaanssen (2007)	Mexico	3	3 months	Landsat 30m	Irrigated wheat	0-500m	Eddy correlation	9%
Teixeira et al. (2009)	Brazil	10	7 years	Landsat 30m	Tree crops	0-500m	SEBAL parameters	-
Kongo et al. (2011)	South Africa	28	4 months	MODIS 1km	Forest, pastures, water bodies	400 - 3,000m	Scintillometer	26%
Sun et al. (2011)	China	1	1 day	Landsat 30m	Lake, Wetlands	40-258m	E-Pan	11%
Ruhoff et al. (2012)	Brazil	28	12 months	MODIS Terra 1km	Sugarcane	500 - 1,500 m	Eddy correlation	9%

Previous research using SEBAL has indeed shown great potential of applying remote sensing to estimate *ET* on few or specific land use types for a limited period of time or with a low temporal resolution. The Upper Pangani River Basin with an elevation range between 600 - 5,900 masl has a higher heterogeneity. It consists of 16 land use types that include snow/ice, forest, irrigated croplands, rainfed agriculture, natural vegetation and water bodies (wetlands, lakes and reservoirs) (Kiptala et al., 2013a).

The high elevation range also influences the inter-seasonal and intra-seasonal ET fluxes for various land use types under different climate conditions. An accurate estimation of ET fluxes is certainly crucial for water resource planning in this river basin.

The SEBAL algorithm was therefore used to map ET fluxes for three consecutive years, i.e. 2008 (wet), 2009 (dry) and 2010 (average). MODIS (Aqua and Terra) data of moderate resolution were utilized. The timestep of 8-day and spatial scale of 250-m were limited by the available MODIS vegetation satellite product. The timescale (8-day) generally corresponds to the time scale that characterizes agricultural water use while 250-m scale is reasonably representative of the sizes of the small-scale irrigation schemes in the Upper Pangani River Basin. Since there are no ET measurements in the basin, the SEBAL results were validated by various proxies that include pan evaporation, reservoir water balance, crop water coefficients and catchment water balance. The SEBAL ET results are also compared with independently computed global ET products. The product chosen is derived from the MODIS 16 algorithm (Mu et al., 2007; 2011) that provides baseline global ET on vegetated land surface at 1 km resolution. The other global ET products have high spatial resolutions and have not been considered. They include: PCR-GLOBWB (Van Beek and Bierkens, 2009), global ET computed at a resolution of 0.5° (56 km) using water balance approach, ERA-Land (Balsamo et al., 2011) and ERA-Interim (Dee et al., 2011) global ET computed at 0.7° (78 km) using land surface model and GLEAM (Miralles et al., 2011) global ET computed at 0.25° (28 km) using remote sensed land surface model.

4.2 MATERIALS AND METHODS

The following section describes the three main datasets for the SEBAL calculations including the pre-processing of the MODIS images. The SEBAL algorithm, MODIS 16 algorithm and in-situ validation methods and the uncertainty assessment are also described in detail.

4.2.1 Datasets

Pre-processing of MODIS datasets

The Moderate-resolution Imaging Spectroradiometer (MODIS) is an extensive program using sensors on two satellites (Terra and Aqua) to provide a comprehensive series of global observations of the Earth's land, oceans, and atmosphere in the visible and infrared regions of the spectrum. Terra earth observation system (EOS) was launched in 1999 while Aqua EOS was launched in 2002. The time of overpass of Terra (EOS AM) satellite is 10.30a.m while Aqua (EOS PM) satellite is 13.30pm local time. The MODIS data is available in different versions, and the latest version 5 (V005) available from 2008 from the USGS database has been validated (USGS, 2012). The images were retrieved from the Land Processes Distributed Active Archive Center (LPDAAC) of the National Aeronautics Space Administration (NASA) [https://reverb.echo.nasa.gov/reverb]. The MODIS images required for the SEBAL model include land surface temperature (LST)/emmissivity (EMM), surface reflectance (SF) and vegetation index (VI) (Table 4.2).

Vegetation Index (VI) products are scaled by multiplying with 0.0001 to provide the Normalized Difference Vegetation Index (NDVI). NDVI is the key (and undisputed) indicator of ET fluxes (Bastiaanssen et al., 2012; Nagler et al., 2005; Burke et al., 2001). The two 16-day NDVI datasets (MOD13 and MYD13) starting on day 1 and day 9 at 250-m were used to create 8-day 250-m NDVI layers. The other MODIS products were therefore acquired and re-projected to this scale for the period 2008 - 2010. The average emissivity (Em) was computed as the average of Em_31 (from band 31) and Em_32 (from band 32) and scaled by 0.002 with a minimum Em of +0.49. Surface reflectance (bands 1 - 7) were also extracted from the daily land surface reflectance products and scaled by 0.0001. Liang's method (Liang, 2001) was used to calculate the broadband surface albedo from the seven surface reflectance bands. Further information on the products is available on the USGS website (USGS, 2012).

Table 4.2: MODIS satellite images used in the SEBAL analysis.

Satellite Imagery	Product/Sensor	Spatial Scale	Temporal scale
Land surface temperature/emissivity.	MOD11_L2 (Terra) & MYD11_L2 (Aqua)	1-km	Daily
Surface reflectance	MOD09GA (Terra) & MYD09GA(Aqua)	500-m	Daily
Vegetative Index (NDVI)	MOD13 (Terra) & MYD13 (Aqua)	250-m	16-day

In total, 138 sets of MODIS images were re-projected to cover the period 2008 - 2010 over the Upper Pangani River Basin. To have continuous satellite data, clouded pixels in the images have to be corrected to minimize uncertainties generally associated with satellite data (Courault et al., 2005; Hong et al., 2009). Clouded pixels were removed and corrected using advanced interpolation techniques in ERDAS imagine software (ERDAS, 2010). For each image with cloud pixels, an area of interest (AOI) was created over the clouded area (only the section of the image with cloud cover). If the AOI is not completely covered by the clouds, the pixels that have correct spectral values were randomly picked and interpolated over the AOI. The AOI size for a particular interpolation is limited to one land use type to ensure that the AOI does not span wide topographical range. If the AOI is fully clouded or large (spans between land use types), the histogram matching option was used to match data with the nearest reliable value (assumed to have similar spectral characteristics) from the next or previously available image. The procedure is similar to the method proposed by Zhao et al. (2005) and also used by MODIS 16 algorithm (see section 4.3.3) to generate continuous global ET which entailed identification and replacement of unreliable pixel value (cloud contaminated) with the nearest reliable value prior to or after the missing data point.

The procedure for cloud removal is critical for Upper Pangani River Basin where most of the clouded pixels occur in the mountainous areas. As such, the uncertainties associated with the interpolation are more pronounced in the mountainous areas. However, we argue that the instantaneous ET does not vary significantly within land use type, e.g. snow, afro-alpine forest that are dominant in the upper catchments (especially during the wet seasons). Furthermore, the model results are scaled using the potential evaporation derived from ground information.

Precipitation datasets

Daily rainfall data for 93 stations located in or near the Upper Pangani River Basin were obtained from the Tanzania Meteorological Agency and the Kenya Meteorological Department. The data was subjected to screening and checked for stationarity and missing data. Of the original group, 43 stations were selected for computing the areal rainfall in the river basin. The selected stations were based on the availability and reliability of the rainfall data for the period of analysis, 2008 - 2010.

Unfortunately, there are no rainfall stations at elevations higher than 2,000 m a.s.l. where the highest rainfall actually occurs. Remote-sensed sources of rainfall data based on or scaled by ground measurements have similar shortcoming, e.g. FEWS and TRMM. According to PWBO/IUCN (2006), the maximum mean annual precipitation (MAP) at the Pangani River Basin is estimated at 3,453 mm yr^{-1} that is estimated to occur at elevation 2,453 m.a.s.l. Therefore, a linear extrapolation method based on the concept of double mass curve was used to derive the rainfall up to the mountain peaks using the rainfall data from the neighbouring stations. It was assumed that the MAP is constant above this elevation to 4,565 m.a.s.l. for Mt. Meru and 5,880 m.a.s.l. for Mt. Kilimanjaro. This assumption is expected to have negligible effect at the Pangani River Basin because of the relative small area above this elevation. Six dummy stations were therefore extrapolated from the existing rainfall stations to the mountain peaks. The rainfall point measurements (including the extrapolated points) were interpolated using the inverse distance method (using ArcGIS Geostatistical Analyst) to develop spatial distribution of rainfall for the Upper Pangani River Basin for year 2008 - 2010 (Fig. 4.1).

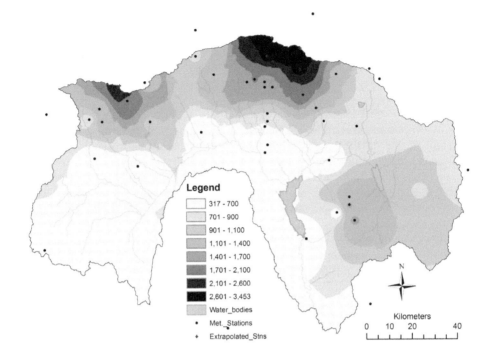

Fig. 4.1: Mean annual precipitation (mm yr⁻¹) for the Upper Pangani River Basin for year 2008 – 2010.

Land use and land cover types

The land use and land cover map for the Upper Pangani River Basin developed in Chapter 3 was used in this chapter.

4.2.2   Surface Energy Balance Algorithm of Land (SEBAL) algorithm.

SEBAL is an energy partitioning algorithm over the land surface, which was developed to estimate (actual) ET from satellite images (Bastiaanssen et al., 1998a; 1998b). SEBAL calculates ET at the time of satellite overpass as a residual term of the surface energy balance. The parameterization is an iterative and feedback based procedure and a detailed description of the SEBAL steps and its applications can be found in Mohamed et al. (2004) and is also available on the Waterwatch website (www.waterwatch.nl). The SEBAL algorithm has been scripted for auto-processing in ERDAS Imagine 9.2 software.

SEBAL estimates the spatial variation of the hydrometeorological parameters of LULC types using satellite spectral measurements and limited ground meteorological data. These parameters are used to assess the surface energy balance terms, which are responsible for the re-distribution of moisture and heat in soil and atmosphere. ET is derived in terms of instantaneous latent heat flux, $\lambda\,E$ (W m⁻²).

$$\lambda E = R_n - H - G \qquad (4.1)$$

Where R_n is the net radiation (W m^{-2}), H is the sensible heat flux (W m^{-2}) and G is the soil heat flux (W m^{-2}). Eq. 4.1 can be expressed as latent heat flux by considering evaporative fraction Λ (-) and the net available energy $(R_n - G_o)$.

$$\Lambda = \frac{\lambda E}{\lambda E + H} = \frac{\lambda E}{R_n - G_o} \qquad (4.2)$$

The daily evapotranspiration is determined by assuming that the evaporative fraction is constant during daytime hours. Shuttleworth et al. (1989) and Nichols and Cuenca (1993) have shown that midday evaporative fraction is nearly equal to average daytime evaporative fraction. Peng et al. (2013) on a recent study of a wide range of ecosystems and climates also established that instantaneous evaporative fraction could represent daytime evaporative fraction especially between 11.00hrs to 14.00hrs local time. Since the overpass time for the satellite images (10.30am and 1.30pm) are reasonably close or within the midday times, this assumption is valid for this study. The validity of this assumption has now been widely adopted by various remote sensing algorithms computing ET over larger scales (Su, 2002; Muthuwatta and Ahmad, 2010; McCabe and Wood, 2006).

The soil heat flux, G represents the heat energy passed through to the soil. G is a small component of the surface energy component relative to the other terms in Eq. 4.1. It is usually positive when the soil is warming and negative when it is cooling. For the time scales of 1 day, G can be ignored (night and day balance) and the net available energy $(R_n - G_o)$ reduces to net radiation (R_n). The assumption of negligible G is also valid at seasonal scale in the tropical climate, since G is not expected to vary significantly. This is unlike the Arctic regions where large portion of G is used to melt ice in the spring to early summer season (Engstrom et al., 2006).

Following these assumptions at the daily timescale, ET_{24} (mm d^{-1}) can be computed using the approach of Bastiaanssen et al. (2002):

$$ET_{24} = \frac{86400x10^3}{\lambda \rho_\omega} \Lambda R_{n24} \qquad (4.3)$$

Where R_{n24} (W m^{-2}) is the 24-h averaged net radiation, λ (2.47 x 10^6 J kg^{-1}) is the latent heat of vaporization and ρ_ω (1000 kg m^{-3}) is the density of water.

The daily ET_{24} has been scaled up to 8-day time scale steps (ET_{8day}) assuming the same proportion variability of potential evaporation ET_o between 1-day to 8-day period (Eq. 4.4). In other words, the ratio of ET_o derived from standard meteorological measurements has been used to represent weather change between the two time steps (Morse et al., 2000).

$$ET_{8day} = (ET_{24}) \times \left(\frac{ET_{o-8day}}{ET_{o-day}} \right) \qquad (4.4)$$

The monthly ET_{month} is the summation of the ET_{8day} for each month.

It is noteworthy that the SEBAL model has a tendency to overestimate λE due to differing extreme pixels (wet and dry) selected by the operator (Long and Singh,

2012; Ruhoff et al., 2012). It is therefore desirable that the users have adequate knowledge and experience on the selection of these pixels in the SEBAL model.

4.2.3 MODIS 16 ET Algorithm

MODIS 16 algorithm (Mu et al., 2007; 2011) computes global ET over vegetated land areas at 1-km, 8-day scales and are available from January 2000. The MODIS 16 algorithm utilizes global MODIS and global meteorology from GMAO (Global Modelling and Assimilation Office - NASA) ground-based meteorological data. MOD 16 algorithms (Mu et al., 2007; 2011) are a revision of an earlier algorithm proposed by Cleugh et al. (2007) based on the Penman-Monteith (P-M) equation (Monteith, 1965):

$$\lambda E = \frac{s \times A + \rho \times C_p \times (e_{sat} - e)/r_a}{s + \gamma \times (1 + r_s/r_a)}$$

(4.5)

Where: $s = d(e_{sat})/dT$ (Pa K^{-1}) is the slope of the curve relating saturated water pressure (e_{sat} (Pa)) to temperature; e (Pa) is the actual water vapor pressure; A (W m^{-2}) is available energy partitioned between sensible heat, latent heat and soil heat fluxes on a land surface; ρ (kg m^{-3}) is the air density; C_p (J Kg^{-1} K^{-1}) is the specific heat capacity of air; γ is psychrometric constant (Maidment, 1993); r_a (s m^{-1}) is the aerodynamic resistance and r_s (s m^{-1}) is surface resistance which is the effective resistance to evaporation from the land surface and transpiration from the plant canopy.

Mu et al. (2007) revised the P-M model by incorporating a soil evaporation component by adding vapor pressure deficit and minimum air temperature constraints on stomatal conductance and upscaling canopy conductance using leaf area index. The input data includes the MODIS data: 1) global land cover (MOD12Q1) (Friedl et al., 2002); 2) Fraction of Absorbed Photosynthetically Active Radiation/Leaf Area Index (FPAR/LAI (MOD15A2)) (Myneni et al., 2002); and 3) MODIS albedo (MCD43B2/B3) (Lucht et al., 2000; Jin et al., 2003). The input non-satellite data are NASA's MERRA GMAO (GEOS-5) daily meteorological data at 1.00°x 1.25° resolution. Cloud-contaminated or missing data are filled in MODIS 16 algorithm at each pixel by a process which entailed identification and replacement of the unreliable pixel value with nearest reliable values prior to and after the missing data point (Mu et al., 2011). The procedure similar to the one proposed by Zhao et al. (2005) to generate continuous global terrestrial ET data on 8-day 1-km scales. The procedure is also similar to the one adopted for this study, however using a pixel instead of an AOI.

However, the initial MODIS 16 algorithm (Mu et al., 2007) significantly underestimated global ET (45.8x10^3 km^3) compared to other reported estimates (65.5x10^3 km^3). The algorithm was further improved by: 1) inclusion of ET as sum of both daytime and night time components; 2) separation of the canopy into wet and dry surfaces; 3) separation of soil surfaces into saturated wet surface and moist surface; 4) estimation of the soil heat flux as radiation partitioned on the ground surface; and 5) improvement of estimates of stomatal conductance, aerodynamic resistance and boundary layer resistance (Mu et al., 2011). The improved MODIS 16 algorithm provided a better estimate of global annual ET over vegetated land namely $62.8 \times 10^3\,\mathrm{km}^3$. Limited validation using eddy flux towers: 46 Ameriflux in the US (Mu et al., 2011) and 17 flux towers in continental to arid climate in Asia (Kim et al., 2011) also showed enhanced global ET results with MAE of below 30% to the measured ET. The MODIS 16 algorithm was observed to provide baseline global ET fluxes for various landscapes on regional and global water cycles (Mu et al., 2007; 2011; Kim et al., 2011).

4.2.4 In-situ ET assessment methods

Since there are no direct measurements of ET using specialized techniques such as Scintillometers or the flux towers (commonly used to validate ET (Table 4.1)) in the studied basin, the study infers other in-situ measurements to assess the accuracy of SEBAL ET fluxes.

Open water evaporation from pan evaporation measurements

Open water evaporation from pan measurements $(E_{o(p)})$ can be estimated from pan evaporation (E_p). E_p records the amount of water evaporated from a pan filled with unlimited supply of water during a day (mm d^{-1}). A class A pan, screened (Allen et al., 1998) is located at the NyM Met Station close to the dam outlet (0.5km to dam, +16m elevation diff. to the reservoir). Since the pan conditions (such as heat storage and transfer, air temperature and humidity, wind conditions) may not be similar to the open water evaporation in the reservoir, the E_p are corrected by pan coefficient factor, K_p to compute $E_{o(p)}$ estimates for the NyM reservoir (Eq. 4.6).

$$E_{o(p)} = K_p \times E_p \qquad\qquad (4.6)$$

K_p ranges between 0.90 - 1.05 for class A pan under moderate wind conditions in tropical climates (Doorenbos and Pruitt, 1977). However, previous studies (e.g. Hoy and Stephens, 1979; Howell et al., 1983; Abtew, 2001) and a recent review article by McMahon et al. (2013) have shown that pan evaporation in semi-arid climates is much higher than open water measurements, with pan coefficient mostly in the range of between 0.7 - 0.9. The higher pan evaporation is attributed to difference in heat conduction between the boundary layers of the water body compared to the pan. However, if the pan has a screen covering (like the case in this study), there is a slight reduction in evaporation attributed to radiation interception by the screen (steel mesh) thus slightly increasing the pan coefficient by around 10% (Howell et al., 1983). It is clear that the pan coefficient is specific to pan, location and nature of the water body (size and depth). In view of this, a pan coefficient of 0.9 is adopted initially for this study and thereafter, an ideal pan coefficient is determined.

Water balance at NyM reservoir

A water balance of the NyM reservoir has also been used to validate open water evaporation (Eq. 4.7).

$$E_{o(b)} = (Q_{in} + P) - \left(Q_{out} + \frac{dS}{dt}\right) \tag{4.7}$$

Where $E_{o(b)}$ [mm month^{-1}] is the evaporation rate of the open water surface, Q_{in} [mm month^{-1}] is the inflow into the reservoir, Q_{out} [mm month^{-1}] discharge and dS/dt [mm month^{-1}] is the change in water storage in the reservoir from the water level measurements. $E_{o(b)}$ is compared with the ET of the open water of the reservoir from the SEBAL model.

Crop coefficients, K$_c$.

The seasonal variability of ET can be evaluated through the variation of the crop coefficient, K_c which is the relative evapotranspiration ratio, (Eq. 4.8).

$$K_c = ET / ET_o \tag{4.8}$$

ET is computed using the SEBAL algorithm, while ET_o is derived from the FAO Penman-Monteith formula defined by weather data (Allen et al., 1998). The ET_o was calculated at four climate stations (locations). The SEBAL ET for the dominant land use type at this locations where used to determine the respective K_c values. The computed seasonal variability of K_c values were then compared with the ideal seasonal K_c coefficients, for that specific land use, under similar climatic conditions (Doorenbos and Pruitt, 1977; Allen et al., 1998).

Catchment water balance

The catchment water budget is evaluated based on the estimates of precipitation (P) and SEBAL ET. The contribution of various land use types to the surface outflow (Q_s) at the outlet of the catchment is computed using Eq. 4.9.

$$Q_s = (P - ET) - dS/dt \tag{4.9}$$

The change in storage (dS/dt) is assumed to be negligible or zero for each land use type in the period under consideration (2008 - 2010). If P exceeds the ET then the land use type is a net contributor to the downstream hydrology. If P is less than ET then the land use type consumes additional blue water resources that could otherwise constitute stream flow. For the whole catchment, Q_s (from SEBAL model) is compared with the measured discharge (Q_o) at the outlet (gauging station, 1d8c) of the Upper Pangani River Basin. In this case, the change of storage at the largest water storage, NyM reservoir (water balance, NyM) is taken into consideration.

4.2.5   Uncertainty assessment in SEBAL ET estimates

Non - parametric significance test

ET estimates have a temporal distribution that is influenced by the seasonal variability of potential evaporation and available green and blue water resources. ET estimates for a given land use type may therefore not follow a normal distribution in time. Large topographic range on a land use type may also influence the distribution of ET values within the same land use type. According to Khan et al. (2006), non-parametric statistical inferences provide more robust results of such data than using classical normal distribution methods. A normality test using the Shapiro-Wilk method (Shapiro and Wilk, 1965) is undertaken as an exploratory test to ascertain the distribution of the ET estimates. Based on the outcome of the exploratory test, two non-parametric tests methods were considered for this study.

First, the most commonly used non-parametric method to test the significance of two estimated means is the Wilcoxon rank sum method (Conover, 1980; Lehmann, 1975). This non-parametric method is used to test the difference of the means of SEBAL ET and MODIS 16 ET estimates presented at monthly scale for all land use types. The other non-parametric method to test the significance of variance of the two estimates is Levene's test (Levene, 1960). The method considers the distances of the ET estimates from their median rather than the mean. Using the median rather than the sample mean makes the test more robust for continuous but not normally distributed data (Levene, 1960; Khan et al., 2006). Both methods use a hypothesis p-value for which the level of significance determines the statistical test. A significance level of 0.05 (confidence level of 95%) is used in the study and if the p-value is greater than 0.05, then one accepts the null hypothesis and if the p-value is less than 0.05 then the null hypothesis is rejected.

Non - parametric confidence interval

The non-parametric bootstrapping technique is used to estimate the confidence intervals in the annual estimates of mean and variance for precipitation (P), ET and effective precipitation (Q_s). The pixel values of P, ET and Q_s for each land use type are used as the sample population or bootstrap sample for the analysis. The average annual values are used to eliminate any potential intra-seasonal variations in the estimates for the period 2008 - 2010. The bootstrapping will draw random samples with replacement from the original population sample each time calculating the mean or variance (Efron and Tibshirani, 1993). The process is repeated 1000 times and a plot of the distribution of the sample means or variance is made. The 95% confidence interval for the mean or variance is determined by finding the 2.5[th] and 97.5[th] percentiles on the constructed distribution. The statistical software Minitab (2003) has been used in determining the bootstrap confidence intervals for the annual estimates of P, ET and Q_s for each land use type.

4.3 RESULTS AND DISCUSSIONS

The monthly ET_{month} calculation is given in Section 4.3.1, computed from the ET_{8day} for 138 time steps covering the years 2008 - 2010. The uncertainty and error assessment of the SEBAL ET results is presented in Section 4.3.2; the seasonal variation of crop coefficient using SEBAL ET data is presented in Section 4.3.3 and the interpretation of the spatio-temporal pattern of water consumption in the Upper Pangani River Basin in Section 4.3.4.

4.3.1 Actual Evapotranspiration

The annual ET results for the Upper Pangani River Basin are given in Fig. 4.2 for the three years of analysis: 2008, 2009 and 2010. The mean annual totals for various LULC types and their monthly variability are given in Figs. 4.3 and 4.4, respectively. The key drivers of the spatial and temporal variability are the dynamics of the precipitation and the biophysical characteristics represented by different LULC types, and the intra/inter-seasonal variation of the climatic conditions in the river basin.

The highest annual ET has been observed for the water bodies and the forested areas. At elevation above 2,300 m.a.s.l, the annual ET values have been gradually reduced by the low atmospheric demand because of low temperatures as the elevation increases. This has also been illustrated by the change in canopy structure of land cover types from dense forest to afro-alpine vegetation and then to the bareland/ice as the elevation increases.

Fig. 4.3 shows the mean annual ET values for different LULC types. It was observed that the annual ET value does not significantly vary with the mean. However, a notable difference has been observed for the LULC in the upper and lower catchments for 2008 and 2009 (Figs. 4.2 & 4.3). For 2008, (a relatively wet year) the annual ET was slightly higher than the mean for the LULC types on the lower catchments (grasslands, shrublands, bushland) due to the enhanced rainfall. However, the annual ET for the LULC types at higher elevations (dense forest, afro-alpine forests) and water bodies was slightly lower because of lower potential ET due to the cooler conditions. Conversely, for 2009 (a relatively dry year), the annual ET for LULC in the lower catchments has been suppressed by limited precipitation but the hotter conditions (higher potential ET) imply higher ET for other LULC types (forest, wetlands, irrigation, water bodies) that have access to additional *blue* water resources (rivers, groundwater).

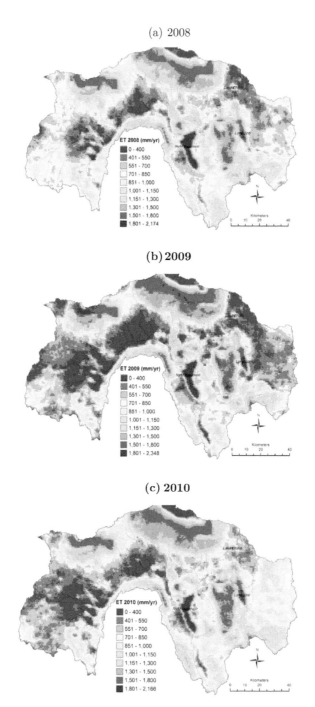

Fig. 4.2: Spatial variation of annual evapotranspiration in the Upper Pangani River Basin for (a) year 2008, (b) year 2009, and (c) year 2010.

Fig. 4.4 shows the temporal variability of mean monthly *ET* for selected LULC types for the period of analysis. The temporal variability has been influenced by the vegetation pattern and the climatic conditions throughout the year. The hotter months of October to March experience generally higher monthly *ET* values while the cooler months from April to July have lower values for all LULC types. Water bodies have higher monthly *ET* values throughout the year, followed by the forest areas and the irrigated croplands. The pastures, shrublands and barelands were found to have the lowest monthly *ET* values. The monthly *ET* values for the bareland/ice were significantly enhanced during the hotter months from October - March when the atmospheric demand (potential evaporation) at the higher altitudes increased.

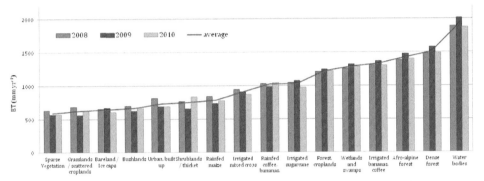

Fig. 4.3: Mean annual evapotranspiration in the Upper Pangani River Basin for different land use types for the years 2008 - 2010.

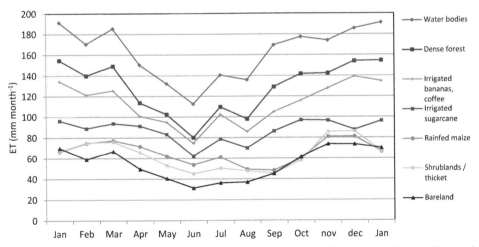

Fig. 4.4: Temporal variation of mean monthly evapotranspiration the Upper Pangani River Basin for selected land use types, averaged over 2008 - 2010.

4.3.2 Model performance

The performance of SEBAL ET estimates were compared with independent ET estimates from MODIS 16 global algorithm and pan evaporation estimates for NyM reservoir. The error analysis was in respect to the correlation coefficient (R), coefficient of determination (R^2), Root Mean Square Error (RMSE) and Mean Absolute Error (MAE) (Table 4.3). The exploratory normality Shapiro - Wilk test resulted in p-values of 0.00 for all ET estimates. The test results, which were below the 0.05 significance level, confirm that the ET estimates do not follow a normal distribution and thus a non-parametric statistical inference is the appropriate method. The non-parametric significance test statistics for mean difference (Wilcoxon) and variance (Levene) for various ET comparisons are also presented in Table 4.3.

Table 4.3: Error statistics and significance test for each validation test using monthly estimates.

Product	Land use type	R	R^2	RMSE mm	MAE mm	Wilcoxon p-value	Levene p-value
MOD 16 ET	vegetated land surface (except water bodies, barelands/ice, urban)	0.74	0.32	28.4	23.9	0.00	0.55
$E_{o(p)}$ - NyM	open water - NyM reservoir	0.95	0.91	8.1	6.3	0.90	0.81

Comparison between SEBAL vs MODIS 16 ET algorithms results

SEBAL ET fluxes were compared specifically with the MODIS 16 ET product to derive any similarity or difference that can inform the model structure or formulation. We note that the SEBAL ET was driven by in-situ meteorological data to generate ET fluxes on 8-day 250-m resolution while MODIS 16 ET was driven by the GMAO meteorological data. MODIS 16 ET only provides ET fluxes for vegetated land surfaces and therefore three land use types; water bodies, bareland/ice and urban were excluded in the analysis. It is noteworthy that the global land-use map used in MODIS 16 ET algorithm is not contemporaneous (geographically) in detail and scale with the land use map (Kiptala et al., 2013a) used in the SEBAL analysis. Therefore, the SEBAL ET land use map was used for statistical analysis to maintain similarity in pixels selection in the evaluation of both ET fluxes. Fig. 4.5 shows the results of the ET comparisons for 13 vegetated land use types at annual and monthly scales.

From Table 4.3, the correlations (at monthly scale) were moderately fair with R of 0.74, R^2 of 0.32, RMSE of 28.4mm month^{-1} (34%) and MAE of 23.9mm month^{-1} (28%). At annual scale the correlation were significantly better with R of 0.91, R^2 of 0.70, and RMSE and MAE of 26% and 24% to SEBAL ET respectively. MAE obtained of 28% on monthly and 24% on annual scales were just within the 10 - 30% range of the accuracy of ET observations (Courault et al., 2005; Kalma et al., 2008; Mu et al., 2011). The regression lines fitted through the origin has a slope of 1.2 in both scales. This implies that the SEBAL ET estimates were 20% more that the

MODIS 16 ET. On monthly (seasonal) scale (Fig. 4.5b), it was observed that SEBAL ET and MODIS 16 ET tends to have better correlations (from 1:1 line) during the cooler months of April, May, June, and July while MODIS 16 ET provided consistently lower ET values during the dry months. The result is also evident from the observations for the dry year 2009 (Fig. 4.5a) that seems to be overestimated compared to the wet (2008) and average (2010) years. The Wilcoxon test result (p-value = 0.00, Table 4.3) shows that the monthly SEBAL ET and MODIS 16 ET means are significantly different at 95% confidence. However, the Levene's test result (p-value = 0.55, Table 4.3) shows that the variances of the two model outputs are statistically the same. Similar significance test results were observed at the annual scale. The test results indicate that the two model results have different means but the same variance. Since the test results for the variance are more robust (Khan et al., 2006), the two model estimates may be considered to be comparable.

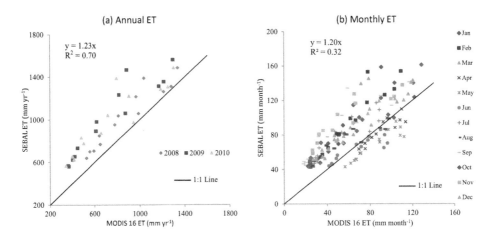

Fig. 4.5: Comparisons of the average SEBAL ET to MODIS 16 ET estimates for different land use types at (a) annual (b) monthly scales for the Period 2008 - 2010 in Upper Pangani River Basin.

From Fig. 4.5, there is a clear trend that MODIS 16 ET estimates are slightly lower than SEBAL ET fluxes during dry periods. It is noted that MODIS 16 algorithm is still undergoing improvement having initially (Mu et al., 2007) underestimated global ET on vegetated land surface. It is notable that the revised algorithm (Mu et al., 2011) provided improved global ET estimates (62.8 x 10^{-3} km³) closer to other reported estimates (65.5 x 10^{-3} km³). However, as observed by Kim et al. (2011), there are still some assumptions inherent in the improved MODIS 16 algorithm such as the stomata closure and zero plant transpiration at night that may result in the underestimation of ET especially during dry periods. Apart from the model structure, high level of uncertainties in the MODIS 16 ET can also be attributed to the coarse resolution of the input data that may be detrimental to ET estimates at a river basin scale. The global land use map used at 1-km may lead to misclassification of certain land uses in such a heterogeneous landscape. This may have lead to biases in the input MODIS FPAR/LAI data in MODIS 16 ET algorithm (Zhao et al., 2006;

Demarty et al., 2007; Mu et al., 2011). Moreover the GMAO meteorological data at 1.0° x 1.25° resolutions is too coarse compared to the ground measurements used in the SEBAL model. It is noteworthy also that the global MODIS ET algorithm (old and new) validation process in North America may also influence the accuracy of the *ET* results in other climatic zones.

Similarly, some assumptions on the estimation of sensible heat flux (*H*) by the SEBAL model if not applied correctly have also been reported to overestimate *ET* especially for dry areas and/or sparse canopy (Mkhwanazi et al., 2012). In estimating sensible heat *H*, most remote sensing approaches make use of radiometric surface temperature instead of aerodynamic temperature (which is difficult to estimate or measure). In doing so, SEBAL in particular introduces a temperature difference gradient that relies on two anchor pixels (wet/cold and dry/hot). The subjective determination of these pixels (despite many recommendations) by the users may introduce uncertainties to the model results. Other SEBAL model assumptions such as the omission of night net radiation (R_n) when it becomes effectively negative or the assumptions that daily heat flux (*G*) is zero can also lead to uncertainties in *ET* estimates (Ruhoff et al., 2012).

Open water evaporation at NyM reservoir

The monthly SEBAL estimates of the open water evaporation ($E_{w(s)}$) at NyM reservoir showed good correlation with *R* of 0.95 and R^2 of 0.91 to pan evaporation estimates (Table 4.3). RMSE values of 8.1mm month^{-1} (5%) and MAE value of 6.3mm month^{-1} (4%) were low, indicating good accuracy between the datasets. However, $E_{p\text{-}NyM1}$ ($K_p = 0.9$) showed a general pattern of overestimation of SEBAL ET by nearly 10% (Fig. 4.6). A review of K_p (to have a linear (1:1) relation) between the *ET* estimates ($E_{p\text{-}NyM2}$) resulted in a reduced K_p factor of 0.81. The pan coefficient (0.81) is reasonable, considering that the site is located on the lower end of the reservoir (0.5km to dam, +16m elevation diff. to the reservoir). The site is also located in a dry environment that is generally associated with lower K_p values. The statistical test for the two *ET* estimate (using $K_p = 0.9$ and $K_p = 0.81$) showed *p*-values greater than 0.05 (Table 4.3) which indicates that both results were not significantly different to the SEBAL estimates at 95% confidence level.

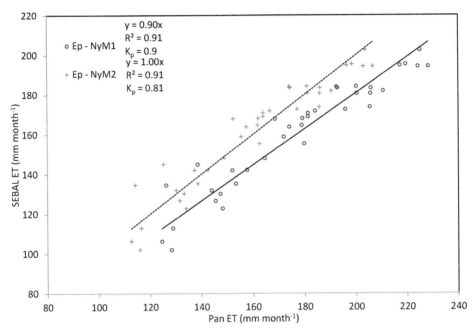

Fig. 4.6: Comparison of SEBAL ET monthly estimates and Pan Evaporation for open water at NyM reservoir for the period 2008 - 2010.

Water balance calculations at NyM reservoir

The open water evaporation at NyM reservoir was also validated through monthly water balance analysis taking into consideration the monthly precipitation, inflows, outflows and changes in water levels (for storage variations) in the reservoir. The total inflows (Q_{in}) and outflows (Q_{out}) were obtained from gauging stations located upstream and downstream of the dam. The precipitation (P) and water level measurements were also obtained from the NyM Met Station and the Pangani Basin Water Office (PBWO). The water levels were also used to compute the surface area of the reservoir at various time steps using formulae adopted from Moges (2003). Table 4.4 shows the annual estimates for each of the water balance components, aggregated from monthly totals, for each year of analysis.

Table 4.4: Annual mean variations of the water balance (mm yr^{-1}) in NyM reservoir for period 2008 – 2010.

	Rainfall (P)	Inflows (Q_{in})	Outflows (Q_{out})	Change in storage (dS/dt)	Evaporation $E_{o(b)}$	% Relative Error to SEBAL ET
2008	385	8479	7355	-631	2139	2
2009	173	5627	6139	-2859	2520	-7
2010	404	7951	5716	728	1912	12

Table 4.4 shows that the relative error (RE) ranged between -7% to +12%. The variations in the RE can be attributed to the measured water levels that may result in high uncertainties in water storage from a relatively shallow dam (active depth of 9m). Nevertheless, the errors even out over the study period with an overall bias of - 2%. The negative RE means the $E_{o(b)}$ from the water balance analysis was slightly lower than the SEBAL ET.

4.3.3 Crop coefficient, K_c for the main crops

Fig. 4.7 shows the K_c (ET/ET_o) seasonal variations computed for four locations under different land use type in Upper Pangani River Basin. The Lyamungu station (Fig. 4.7a) is the most upstream station where irrigated bananas, coffee intercropped with maize and beans are dominant land use. The agricultural activity is intensive throughout the year due to the availability of additional blue water resources. K_c values at this station were greater than 1.0 experienced mostly throughout the period of analysis. The results are consistent with the ideal K_c values for such crops ranging between 1.05 - 1.2 (without water stress) (Allen et al., 1998). However, the climatic conditions, cropping calendar of the intercropped cereals and the type of irrigation used (traditional furrow system) might have contributed to K_c values (greater than 1.2) in some months in wet seasons and similarly lower K_c values (below 1.0) in few months in dry seasons. In 2009 (dry year), the K_c values for month of Jan - Mar (dry season) were much lower due to the water stress from the drought conditions experienced during that year.

The TPC station is located within the TPC sugarcane plantation at the lower catchments of the Upper Pangani River Basin. The cropping calendar of the sugarcane plantation has been designed for continuous sugarcane harvesting (of near equal quantity) between June - February every year. During the long rains (*Masika* seasons from March to May) there is no irrigation to allow for maintenance works at the canals. The crop calendar is therefore designed to ensure that the sugarcane is at different stages of development making use of precipitation. K_c (without water stress) for irrigated sugarcane ranges from 0.4 - 1.25 for homogenous sugarcane plantation with continuous cropping stages (Allen et al., 1998). However, since the cropping stages were mixed, the ideal (mean) K_c would be approx. 0.8 with slightly higher values during the *Masika* season when the all sugarcane is at different stages of maturity. The computed K_c values for irrigated sugarcane (Fig. 4.7b) varied slightly but within the ideal value of 0.8. The K_c values were slightly higher than 0.8 in the *Masika* seasons apart from year 2008. The year 2008 (wet) experienced suppressed rainfall in the month of April compared to subsequent high rainfall in the other months. During the dry months, the K_c values were lower than expected mean (0.8) and were more pronounced during dry year (2009). This result can be attributed to the water stress conditions for the sugarcane due to limited precipitation (*Masika* season) or inadequate water supply for irrigation in dry months.

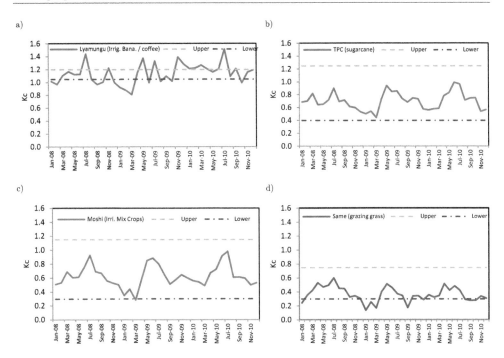

Fig. 4.7: Seasonal variation of ET/ET$_o$ (K_c) at locations: a) Lyamungu b) TPC c) Moshi and d) Same in Upper Pangani River Basin for the years 2008 - 2010.

Moshi station (Fig. 4.7c) is located in the middle catchment, where mixed cereals (maize, beans) and few vegetable crops is dominant land use practice under supplementary irrigation. The agricultural activities rely on rainfall and supplementary irrigation during the wet seasons. The K_c values would therefore be related to the seasonal rainfall and cropping patterns in the areas. The K_c for this station was observed to be high between the months of March and August during the crop growing season and low during the dry months of between September and February. The K_c ranges between 0.3 and 1.0 which was reasonable within the ranges for maize and vegetable crops (0.30 - 1.15) (Allen et al., 1998).

Same station (Fig. 4.7d) is located on the lower catchments with low precipitation (500 mm yr^{-1}) and is dominated by grasslands (for grazing) and scattered croplands. Due to the very dry conditions in this area, the grassland experiences water stress and this is likely the reason why the calculated K_c values are lower than the reported K_c for grazing pasture that range between 0.30 to maximum of 0.75 (Doorenbos and Pruitt, 1977; Allen et al., 1998). The K_c values calculated for this LULC type ranged from 0.2 during the dry seasons and 0.6 during the wet seasons.

4.3.4 Spatio-temporal pattern of water use and catchment water balance

Given the precipitation (P) and the SEBAL ET results, the net contribution or consumption of surface outflow (Q_s) was evaluated for each LULC type (without surface/reservoir storage change) using simple water budget (Eq. 4.9). The usability and reliability of Q_s for water resource planning depends on the confidence intervals *(CI)* of P and ET estimates. The uncertainty of the LULC map is assumed to be inherent on the statistical estimates for each land use type. The lower and the upper bound confidence levels were estimated at 95% confidence limits. Since there was a minimal difference between the upper and lower *CI* (Fig. 4.8) an average *CI* were used and presented in Table 4.5.

Table 4.5: Annual variations of the water balance terms in Upper Pangani River Basin for period 2008 - 2010.

No.	Land use and land cover	km²	Mean Annual P (mm/yr) Mean	STDEV	C.I	Mean Annual ET (mm/yr) Mean	STDEV	C.I	Q (mm/yr) Mean	C.I
1	Water bodies	100	603	82	4	1,928	204	10	*-1,325*	14
2	Bareland/ice caps	100	2,196	612	30	643	653	32	*1,553*	62
3	Sparse vegetation	445	714	301	7	586	172	4	128	11
4	Bushlands	1,152	831	312	5	669	312	5	162	9
5	Grasslands/scattered crops	1,517	691	159	2	630	223	3	61	5
6	Shrublands/thicket	3,509	785	151	1	756	85	1	29	2
7	Rainfed maize	2,942	785	221	2	789	221	2	-4	4
8	Afro-alpine forest	257	2,300	322	10	1,429	309	9	*871*	19
9	Irrigated mixed crops	598	888	324	7	905	207	4	-17	11
10	Rainfed coffee/irrig. banana	723	1,026	250	5	1,022	261	5	5	9
11	Irrigated sugarcane	89	572	204	11	1,035	212	11	*-463*	22
12	Forest, irrig. croplands	556	1,115	366	8	1,228	250	5	-113	13
13	Irrigated bananas, coffee	607	1,449	297	6	1,330	156	3	119	9
14	Dense forest	637	1,703	324	6	1,517	144	3	186	9
15	Wetlands and swamps	98	644	127	6	1,291	267	13	*-647*	20
16	Urban, built up	8	977	117	20	774	80	14	202	34
	Total	13,337	917		4	866		3	52	7

The *CI* (uncertainty of the estimates) of the water balance terms is influenced greatly by the spatial coverage and the distribution range of the land use types. For individual land use types, the *CI* for P and ET ranged between 1 and 3 mm yr[-1] (less than 1%) for the dominant land use types e.g. grasslands, shrublands, and rainfed maize. For land use types of lower spatial coverage *CI* ranges for P and ET were marginally higher with bareland having the highest uncertainty of 32 mm yr[-1] (5%) for ET estimates. The *CI* values for the surface outflow, Q_s were the accumulated totals *CI* for P and ET. For the entire catchment, the uncertainty of the mean estimates of P and ET was low at 3 - 4 mm yr[-1] (less than 1%). However, the cumulative uncertainty for Q_s was higher at 7 mm yr[-1] (13% to the mean of Q_s).

Irrigated sugarcane, wetlands & swamps and the water bodies were found to be the highest net evaporative water users with a consumption of -463 ($\pm$22) mm yr[-1], -647 ($\pm$20) mm yr[-1] and -1,325 ($\pm$14) mm yr[-1] respectively. The afro-alpine forest and bareland/ice caps were the lowest water users contributing downstream flow in excess of 871 ($\pm$19) mm yr[-1] and 1,553 ($\pm$62) mm yr[-1] of the annual precipitation. The total evaporative water use, 866 mm yr[-1], thus accounts for 94% of the annual precipitation in the Upper Pangani River Basin with the remainder of about 52 ($\pm$7) mm yr[-1] or 21

(±2) $m^3 s^{-1}$) estimated to flow to the Lower Pangani River Basin. However, this result will have to be adjusted slightly to account for changes in storage in NyM reservoir regulate flow (artificially) downstream for the period of analysis (approx. $-3.2m^3 s^{-1}$ from Table 4.4). The change in storage was initially assumed to be negligible for various LULC types. This provides an estimated surface outflow of 18 (±2) $m^3 s^{-1}$ which compares reasonably well with the measured outflow (at gauge 1d8c below NyM reservoir) of $20.5m^3 s^{-1}$ (12% bias) for the same period. The bias or error (12%) is within the uncertainty range Q_s estimates of 13% (7mm yr^{-1}).

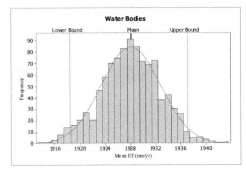

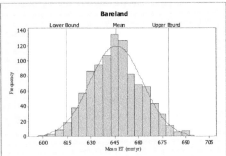

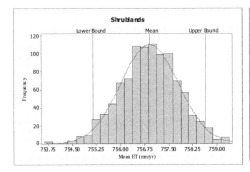

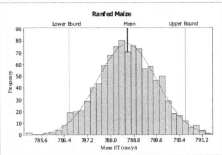

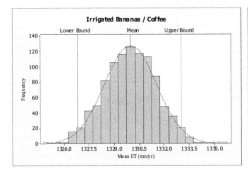

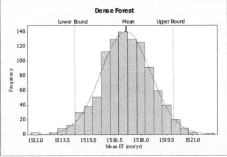

Fig. 4.8. Frequency distribution of the estimated annual SEBAL ET from bootstrap for selected land use types in the Upper Pangani River Basin for period 2008 - 2010.

The result is also consistent with previous analyses of outflows at NyM reservoir which estimated flows of between 15 - 30 m^3 s^{-1} based on long term discharge measurements (Turpie et al., 2003; Komakech et al., 2011; Notter et al., 2012). According to PBWO/IUCN (2006), the hydropower commitments (which exist as a water right since the 1970) for the hydropower production at NyM HEP is 760 Million m^3 yr^{-1} (or about 24 m^3 s^{-1}). The downstream flow is also meant to regulate flow to Hale HEP and the (new) Pangani HEP (Fig. 2.1). Considering these HEP flow commitments, notwithstanding the irrigation water needs and the environmental flow requirements for the Lower Pangani River Basin, the Upper Pangani River Basin is indeed a closed or closing basin (considering the uncertainties), with its river systems under stress (Molden et al., 2005; Molle et al., 2005).

4.4 CONCLUSION

This research has used MODIS data and the SEBAL algorithm to estimate spatio-temporal ET in a data scarce river basin in Eastern Africa with a highly heterogeneous use of water. A good agreement was generally attained for the SEBAL ET results from the various validations. For open water evaporation, the SEBAL ET for NyM reservoir, showed a good correlation with the pan evaporation measurements using an optimized pan coefficient of 0.81. Similarly, the water balance ET estimates for NyM reservoir resulted in an absolute relative error 2% on the mean annual estimates over the study period. The estimated ET for various agricultural land uses indicated a pattern that was consistent with the seasonal variability of the crop coefficient (K_c) based on FAO Penman-Monteith equation. As expected, ET estimates for the mountainous areas experiencing afro-alpine climate conditions have been significantly suppressed by the low potential ET. For the whole basin, ET accounted for 94% of the total precipitation with a surface outflow closure difference of 12% to the measured discharge. The bias range (12%) was within the uncertainty (13%) level at 95% confidence interval for P-ET estimates.

Comparison between global MODIS 16 ET and SEBAL ET showed good correlation R of 0.74. However, the R^2 was lower at 0.32 and the RMSE and MAE where 34% and 28% respectively, with the MAE being just within the acceptable comparison level of below 30%. The monthly ET variance of the two models was not statistically different whereas the monthly ET mean was statistically different. In general, the MODIS 16 ET underestimated the SEBAL ET by approximately 20%, mostly during the dry month or seasons. This difference can be attributed to the model structure and the coarse spatial scale of the MODIS 16 ET. The difference might also have been exacerbated by SEBAL's tendency of overestimating ET in dry periods.

The study has established that the ET during a relatively dry year (2009) is higher for LULC in the upstream catchment, such as forests and irrigated croplands, due to the local availability of blue water resource (from snow melts, rivers and groundwater). ET for water bodies (lakes and reservoirs) and irrigated croplands that extract water from the river systems is also higher. However, for LULC types that have limited access to blue water, such as rainfed agriculture and grasslands, the ET is lower due to the limited precipitation. Conversely, in a relatively wet year (2008), the ET is suppressed in the upstream catchments due to lower potential evaporation while it is enhanced from the LULC types in the lower catchments due to availability of water resource from precipitation. This result demonstrates the vulnerability of water users in the lower catchments to climate variability and future water scarcity.

This study has highlighted the levels of water use of each LULC type and their relative contribution and/or effect on the downstream hydrology. The water balance approach showed that the basin is closing. A viable option is improving water productivity through improved water efficiency and water re-allocation. The derived spatially distributed ET can provide useful information for a systematic approach of water accounting (Karimi et al., 2013a). The satellite-derived ET fluxes (which also accounts for blue water use) can also provide crucial information for hydrological modelling in highly utilized and water stressed river basins (Winsemius et al., 2008; Zwart et al., 2010; Romaguera et al., 2012).

A major limitation in deriving remote-sensed ET especially for land use types on higher elevations in the humid to sub-humid tropics is the persistent cloud cover. As such, the multi-temporal scales provided by MODIS (Table 4.2) offered a range of images at a reasonable interval (for this case 8-day). These images also enhance the quality of the cloud filling procedure adopted in this study that relies on the next or previous good quality image. This advantage is however limited by the moderate spatial scale of the MODIS images (250-m, 1 km thermal).

Chapter 5

MODELLING STREAM FLOW USING STREAM MODEL[3]

Integrated water resources management is a combination of managing blue and green water resources. Often the main focus is on the blue water resources, as information on spatially distributed evaporative water use is not readily available as is the link to river flows. Physically based spatially distributed models are often used to generate this kind of information. These models require enormous amounts of data, which can result in equifinality, making them less suitable for scenario analyses. Furthermore, hydrological models often focus on natural processes and fail to account for anthropogenic influences. This study presents a spatially distributed hydrological model that has been developed for a heterogeneous, highly utilized and data scarce river basin in Eastern Africa. Using an innovative approach, remote sensing derived evapotranspiration and soil moisture variables for three years were incorporated as input data in the Spatial Tools for River basin Environmental Analysis and Management (STREAM) model. To cater for the extensive irrigation water application, an additional blue water component (Q_b) was incorporated in the STREAM model to quantify irrigation water use. To enhance model parameter identification and calibration, three hydrological landscapes (wetlands, hill-slope and snowmelt) were identified using field data. The model was calibrated against discharge data from five gauging stations and showed a good performance especially in the simulation of low flows where the Nash-Sutcliffe Efficiency of the natural logarithm (E_{ns_ln}) of discharge were greater than 0.6 in both calibration and validation periods. At the outlet, the E_{ns_ln} coefficient was even higher (0.90). During low flows, Q_b consumed nearly 50% of the river flow in the basin. Q_b model result for irrigation was comparable to the field based net irrigation estimates with less than 20% difference. These results show the great potential of developing spatially distributed models that can account for supplementary water use. Such information is important for water resources planning and management in heavily utilized catchment areas. Model flexibility offers the opportunity for continuous model improvement when more data become available.

---

[3] This chapter is based on: Kiptala, J.K., Mul, M.L., Mohamed, Y.A., Van der Zaag, P., 2014. Modelling stream flow and quantifying blue water using modified STREAM model for a heterogeneous, highly utilized and data scarce river basin in Africa. *Hydrology Earth System Sciences*, 18, 2287-2303.

5.1 INTRODUCTION

Hydrological models are indispensable for water resource planning and management at catchment scale as these can provide detailed information on, for example, impacts of different scenarios and trade-off analyses. Society's demand for more accountability in the management of externalities between upstream and the downstream water users has also intensified the need for more predictive and accurate models. However, complexity of hydrological processes and high levels of heterogeneity present considerable challenges in model development. Such challenges have been exacerbated over time by land use changes that have influenced the rainfall partitioning into *green* (soil moisture) and *blue* (runoff) water resources. In spite of these challenges, it is still desirable to develop a distributed hydrological model that can simulate the dominant hydrological processes and take into account the various water uses. In large catchments with high heterogeneity, key variables such as water storage (in unsaturated and saturated zones) and evaporation (including transpiration) are difficult to obtain directly from point measurements. This becomes even more difficult for ungauged or poorly gauged river basins.

In most cases those variables are derived from models using (limited) river discharge data which increases equifinality problems (Savenije, 2001; Uhlenbrook et al., 2004; McDonnell et al., 2007; Immerzeel and Droogers, 2008). On the other hand, grid based distributed models at fine spatial scales do not explicitly account for additional *blue water* use (Q_b), i.e. transpiration from supplementary irrigation or withdrawals from open water evaporation. In fact in tropical arid regions, Q_b can be a large percentage of the river discharge during low flow. Calibrating models using modified stream flow data may lead to incorrect parameterization, and may lead to high predictive uncertainty in the hydrological model outputs especially when dealing with scenarios for water use planning.

To overcome these challenges, many researchers have opted for simple, lumped and or parsimonious models with a limited number of model parameters. The models are simplified by bounding and aggregation of some functionality in the complex system (Winsemius et al., 2008). In doing so, models may become too simplified to represent hydrological processes in a catchment (Savenije, 2010). Therefore, Savenije (2010) proposes a conceptual model mainly based on topographic characteristic to represent the dominant hydrological processes. The model maintains the observable landscape characteristics and requires a limited number of parameters. Other researchers have used secondary data, e.g. from remote sensing to calibrate or infer model parameters as much as possible (Winsemius et al., 2008; Immerzeel and Droogers, 2008; Campo et al., 2006). This has been possible in the recent past because of the availability of satellite images with finer spatial resolutions. Advancement in remote sensing algorithms has also resulted in wider range spatial data of reasonably good accuracies. Such spatial data include actual evapotranspiration (ET_a) derived from remote sensing data, e.g. TSEB (Norman et al., 1995), SEBAL (Bastiaanssen et al., 1998a; 1998b), S-SEBI (Roerink et al., 2000), SEBS (Su, 2002) and METRIC (Allen et al., 2007). Spatial data on soil moisture can also be derived from satellite images, e.g. from ERS-1 Synthetic Aperture Radar (SAR) combined with the TOPMODEL topographic index (Scipal et al., 2005) or from Advance Very High Resolution

Radiometer (AVHRR) combined with the SEBAL model (Mohamed et al., 2004). It is also evident that distributed models perform well with finer resolution data as demonstrated by Shrestha et al. (2007). Using different resolution data (grid precipitation and grid ET_a) and a concept of IC ratio (Input grid data area to Catchment area) they found that a ratio higher than 10 produces a better performance in the Huaihe River Basin and its sub-basin of Wangjiaba and Suiping in China (Shrestha et al., 2007).

Furthermore, remotely sensed data at finer resolutions offer great potential for incorporating blue water, in the form of (supplementary) water use (Q_b) in model conceptualization. This opportunity arises from the fact that remotely sensed ET_a based on energy balance provides total evapotranspiration that already accounts for Q_b. For instance, Romaguera et al. (2012) used the difference between Meteosat Second Generation (MSG) satellites data (total ET_a) and Global Land Data Assimilation System (GLDAS) which does not account for Q_b, to quantify blue water use for croplands in Europe with a reasonable accuracy. However, the spatial scales of such datasets (GLDAS (1 km) and MSG (3 km)) limit the application. Nevertheless, the latter recommended such application to recently available data of wider spatial and temporal coverage, e.g. data derived from Moderate-resolution Imaging Spectroradiometer (MODIS) 250-m, 500-m.

However, the literature shows limited applications of utilizing grid data for distributed hydrological models in poorly gauged catchments. Winsemius et al. (2006) showed that the soil moisture variations from the Gravity Recovery And Climate Experiment (GRACE) could provide useful information to infer and constrain hydrological model parameters in the Zambezi river basin. Campo et al. (2006) using an algorithm developed by Nelder and Mead (1965), used remotely sensing soil moisture information to calibrate a distributed hydrological model in the Arno basin, Italy. Immerzeel and Droogers (2008) used remotely sensed ET_a derived from SEBAL in the calibration of a Soil and Water Assessment Tool (SWAT) model of the Krishna basin in southern India in which the model performance (r^2) increased from 0.40 to 0.81. Recently, Cheema et al. (2014) has used satellite derived rainfall to parameterize the SWAT model while ET_a from ETLook was used to calibrate the model to determine the contribution of groundwater use to the total blue water use in the Indus Basin.

The factors that may have limited the application of remote sensing (RS) data on hydrological modelling include: a) Limited flexibility of hydrological models to utilize spatially distributed data. This is normally the case where the user has no control over the model source code. The user is therefore limited to optimizing model performance using secondary data. b) Limited availability of RS data at the appropriate spatial and temporal scales to capture dominant hydrological processes in a catchment. c) The lack of technical skills by most hydrologists and water resource specialists on how to transform RS data into hydro-meteorological data (Schultz, 1993). The opportunities and challenges for the wider application of remote sensing for hydrological modelling are discussed by De Troch et al. (1996) and Schultz (1993).

This chapter presents a novel method of using ET_a and soil moisture data derived from satellite images as input in a distributed hydrological model. The Upper

Pangani River Basin in Eastern Africa has been used as a case study. This river basin has heavily managed landscapes dominated by small and large scale irrigated agriculture. The secondary data used in this study have been generated using MODIS satellite information and the SEBAL model on 250-m and 8-day resolutions for the period 2008-2010 (Kiptala et al., 2013b). Here the STREAM model has been modified to incorporate blue water use. The model parameters have also been confined further by the topographic characteristics and groundwater observations using the hydrological conceptualization developed by Savenije (2010).

5.2 MATERIALS AND METHODS

5.2.1 Datasets

Hydro-meteorological data

Daily rainfall data for 93 stations located in or near the Upper Pangani River Basin were obtained from the Tanzania Meteorological Agency and the Kenya Meteorological Department. The data was screened and checked for stationarity (Dahmen and Hall, 1990). Of the original group, 43 stations proved useful after data validation for the period 2008 - 2010. Unfortunately, there were no rainfall stations at elevations higher than 2,000 m.a.s.l. where the highest rainfall actually occurs. Spatially distributed rainfall can also be provided by satellite sensors to augment rainfall data from the ground stations (Huffman et al., 2001). Such satellites sensors include the Tropical Rainfall Measuring Mission (TRMM). Famine Early Warning System (FEWS) product also provides remotely sensed rainfall data in Africa. The satellite based rainfall has uncertainties that can be corrected using limited ground rainfall measurements (Hong et al., 2006; Cheema and Bastiaanssen, 2012). Since there were no rainfall stations at the mountainous areas, the satellite based rainfall could not be validated (Haque, 2009).

According to PWBO/IUCN (2006), the maximum long term mean annual precipitation (MAP) at the Pangani River Basin is estimated at 3,453 mm yr^{-1} at elevation 2,453 m.a.s.l. The estimates were based on a rain gauge station that is no longer operational. Therefore, a linear extrapolation method based on the concept of double mass analysis (Wilson, 1983) was used to derive the seasonal rainfall up to the mountain peaks. Double mass analysis assumes relatively consistent correlation between time series of rainfall data at nearby stations with similar hydrological conditions (Chang and Lee, 1974). In the analysis, the seasonal precipitation at the mountain peak (Y) is assumed to have a linear relation to the seasonal precipitation of the nearby stations (X) scaled by a proportionality factor (α). The proportionality factor, α is the average slope of the long term MAP for the two reference points. Y is therefore given as $Y = \alpha X$. The rainfall was maintained constant above this elevation to 4,565 m.a.s.l. for Mt. Meru and 5,880 m.a.s.l. for Mt. Kilimanjaro. This assumption is expected to have negligible effect at the Pangani River Basin because of the relative small area above this elevation (3%). Six dummy stations were therefore extrapolated from the existing rainfall stations to the mountain peaks.

River discharges for six gauging stations were obtained from the Pangani Basin Water Office (Moshi, Tanzania). The measurements were obtained as daily water level measurements and converted to daily discharge data using their corresponding rating curves equations for the period 2008 - 2010.

Evaporation and soil moisture

The actual evapotranspiration (ET_a) and soil moisture data for the Upper Pangani River Basin were obtained from a recent and related research by Kiptala et al. (2013b). ET_a and soil moisture data for 8-day and 250 m resolutions for the years 2008 - 2010 were derived from MODIS satellite images using the Surface Energy Balance Algorithm of Land (SEBAL) algorithm (Bastiaanssen et al., 1998a; 1998b). Actual evapotranspiration (ET_a) is comprised of interception (I), soil evaporation (E_s), open water evaporation (E_o) and transpiration (T).

Land use and land cover types

In this study, we employed the LULC classification for the Upper Pangani River Basin developed by Kiptala et al. (2013a). They derived the LULC types using phenological variability of vegetation for the same period of analysis, 2008 to 2010. LULC types include 16 classes dominated by rainfed maize and shrublands that constitute half of the area in the Upper Pangani River Basin.

Other Spatial data

Elevation and soil data were also obtained for the Upper Pangani River Basin. A digital Elevation Model (DEM) with 90 m resolution was obtained from the Shuttle Radar Topography Mission (SRTM) of the NASA (Farr et al., 2007). The soil map was derived from the harmonized world soil database which relied on soil and terrain (SOTER) regional maps for Northern and Southern Africa (FAO/IIASA/ISRIC/ISS-CAS/JRC, 2012).

5.2.2   Model development

The hydrological model was built to simulate stream flow for the period 2008-2010 for the Upper Pangani River Basin. An 8-day timestep and 250-m moderate resolutions has been used to correspond to availability of remotely sensed ET_a data for the period of analysis. The 8-day time step is sufficiently short for the agricultural water use process, which has a timescale range of between 10 - 30 days (unsaturated zone storage over transpiration rate). In addition, this timescale is assumed to be sufficiently large to neglect travel time lag in the river basin. The other general hydrological processes in the river basin are estimated to have larger time scales (Notter et al., 2012). The spatial scale of 250-m is limited by the available MODIS satellite data. This is reasonably representative of the sizes of the small-scale irrigation schemes in the Upper Pangani River Basin.

STREAM, a physically based conceptual model, was developed in the PcRaster modelling environment (Aerts et al., 1999). The PcRaster scripting model environment consists of a wide range of analytical functions for manipulating Raster

GIS maps (Karssenberg et al., 2001). It uses a dynamic script to analyze hydrological processes in a spatial environment. The PcRaster environment allows for tailored model development and can therefore be used to develop new models, suiting the specific aims of the research including the availability of field data. The STREAM model in PcRaster environment allows the inclusion of spatially variable information like ET_a and soil moisture in the model. Furthermore, STREAM model is an open source model which has been applied successfully in other data limited river basins, especially in Africa (Gerrits, 2005; Winsemius et al., 2006; Abwoga, 2012; Bashange, 2013).

In the STREAM model, surface runoff is computed from the water balance of each individual grid cell, which is then accumulated in the local drainage direction derived from DEM to the outlet point (the gauging station). The model structure consists of a series of reservoirs where the surface flows are routed to the rivers. We modified the STREAM model by including an additional blue water storage parameter (S_b) that regulates Q_b in the unsaturated zone. Q_b can be derived from the groundwater as capillary rise, $C(t)$, or river abstraction, $Q_d(t)$. The input variables for the modified STREAM model are: Precipitation (P), Interception (I) calculated on a daily basis as a pre-processor outside the model. Evaporation (E_s, E_o) and Transpiration (T) denoted as $[E + T]$ was derived by subtracting I for the total evaporation (ET_a) derived from SEBAL $[ET_a - I]$. The minimum soil moisture, $S_{u,min}$ is also derived from SEBAL. The other parameters are determined through calibration. Fig. 5.1 shows the modified STREAM model structure for Upper Pangani River Basin.

In the model $E+T$ and the $S_{u,min}$ are the main drivers of the hydrological processes in the unsaturated zone of the model. $E+T$ is the evaporation (soil moisture) depletion component while $S_{u,min}$ is the depletion threshold. It is assumed that excess water from the upstream cells or pixels would supplement water needs of the middle or lower catchments where supplementary water is used. The Upper Pangani River Basin is a typical river basin, where precipitation exceeds ET_a in the upper catchments and hence contributes river flow to the downstream catchments.

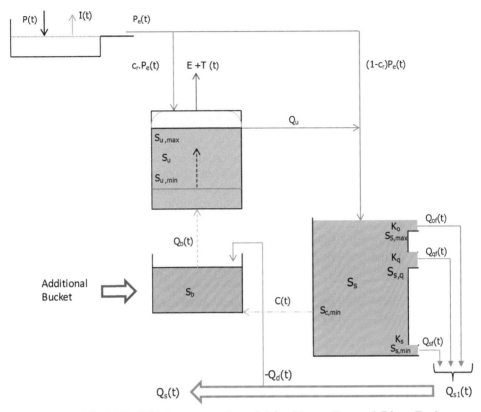

Fig. 5.1: Modified STREAM conceptual model for Upper Pangani River Basin.

The rationale for accounting for Q_b in the model is motivated by the incapability of the original STREAM model if applied in irrigated landscapes to simulate actual transpiration. The original STREAM model was developed specifically for natural landscapes dominated by woody savannas and wetlands with high storage capacity (Dambos) in the Zambezi River Basin (Gerrits, 2005; Winsemius et al., 2006). The blue water use is therefore limited and has been accounted for by the capillary rise only. The total transpiration was therefore derived only as a function of potential evaporation and the soil moisture (from precipitation) in the unsaturated zone using the relation by Rijtema and Aboukhaled (1975). Bashange (2013) using the original STREAM model found that simulated $E + T$ for irrigated croplands were significantly lower compared to SEBAL $E + T$ for dry seasons in the Kakiwe Catchment, Upper Pangani River Basin. The result was attributed to lower soil moisture levels at the unsaturated zone (not replenished in the model by blue water use).

5.2.3   Model configuration

Model input

Interception (I)

When precipitation occurs over a landscape, not all of it infiltrates into the subsurface or becomes runoff. Part of it evaporates back to the atmosphere within the same day the rainfall takes place as interception. The interception consists of several components that include canopy interception, shallow soil interception or fast evaporation from temporary surface storage (Savenije, 2004). The interception is dependent on the land use and is modeled as a threshold value (D). The interception process typically has a daily time scale, although some work has been done to parameterize the interception threshold on a monthly timescale (De Groen and Savenije, 2006).

In our case, we calculate the daily interception according to Savenije, (1997; 2004) outside of the model (see Eq. 5.1);

$$I_d = \min(D_d, P_d) \tag{5.1}$$

where I_d is the daily interception, D_d is the daily interception threshold and P_d is the observed precipitation on a rainy day. Since I_d occurs on a daily time step during a precipitation (P_d) event, the interception at 8-day ($I_{d(8)}$) is derived from the accumulated daily interception computed based on daily precipitation. The interception thresholds (D_d) vary per land use and have been adopted from the guidelines provided by Liu and de Smedt (2004) and Gerrits (2010). As such an interception threshold of 2.5 mm day^{-1} was used for croplands and natural vegetation and 4 mm day^{-1} for forest.

Net Precipitation (Pe)

The net precipitation ($P_{e(8)}$) is calculated by subtracting the accumulated interception ($I_{d(8)}$) from the accumulated precipitation ($P_{d(8)}$) for the 8-day time scale.

$$P_{e(8)} = \sum_{0}^{8} (P_d - I_d) \qquad \forall_t \tag{5.2}$$

$P_{e(8)}$ is split through a separation coefficient (c_r) into the two storages, unsaturated and saturated (groundwater) storages. c_r is a calibration parameter that is dependent on the soil type and land use types.

Evaporation depletion (E + T)

The evaporation depletion ($E + T$) is derived by subtracting the interception component of the actual evapotranspiration (ET_a) at each timestep. ET_a from SEBAL includes $I_{d(8)}$ at 8-day time step.

$$E + T = \left(ET_a - I_{d(8)}\right) \tag{5.3}$$

Unsaturated zone

The maximum soil moisture storage $(S_{u,max})$ was defined based on land use and soil types. Water available for evaporation depletion includes water infiltrated from precipitation $(c_r{\times}P_e)$ and blue water use (Q_b), consisting of water from capillary rise (C) and river abstraction (Q_d). During the dry (nonrainy) periods, the spatial variation in soil moisture is controlled by vegetation through the uptake of blue water resources (Seyfried and Wilcox, 1995). The model assumes a minimum soil moisture level $(S_{u,min})$ which varies for managed and natural landscapes. Soil moisture status at each time step (S_u) is therefore a key variable controlling water and energy fluxes in soils (Eq. 5.4 & 5.5).

$$Q_b = E + T \rightarrow if\left(S_u \leq S_{u,min}\right) \tag{5.4}$$

$$Q_b = 0 \rightarrow if\left(S_u > S_{u,min}\right) \tag{5.5}$$

As a result the green water use is defined as the evaporation depletion less the blue water use (Eq. 5.6).

$$Q_g = E + T - Q_b \tag{5.6}$$

The value for $S_{u,min}$ for each land use type is assumed to be realized during the dry months and is expressed as a fraction of $S_{u,max}$ (soil moisture depletion fraction). $S_{u,min}$ is derived in the SEBAL model for dry months as an empirical function of the evaporative fraction, Λ (the ratio of the actual to the crop evaporative demand when the atmospheric moisture conditions are in equilibrium with the soil moisture conditions) (Ahmed and Bastiaanssen, 2003), see Eq. (5.7).

$$f = \frac{S_{u,min}}{S_{u,max}} = e^{(\Lambda-1)/0.421} \tag{5.7}$$

where f is the soil moisture depletion fraction expressed as a fraction of soil moisture, $S_{u,min}$ to the moisture value at full saturation, $S_{u,max}$ for the dry months. $S_{u,min}$ was realized in the month January which is the driest period in the river basin. Values for f are given in Fig. 5.2 for selected land use types for the dry month of January averaged over 2008-2010.

The soil moisture levels agree reasonably well with previous field studies that have shown similar ranges for natural land use types in sub humid and semi - arid areas (Fu et al., 2003; Korres et al., 2013). It is also noted that the SEBAL model has some level of uncertainty to soil moisture storage and water stress (Ruhoff et al., 2012). In recognizing this uncertainty, the modified SEBAL model also uses a water balance approach where lower $S_{u,min}$ levels can be tolerated with respect to the available Q_b during the dry season for natural land use types.

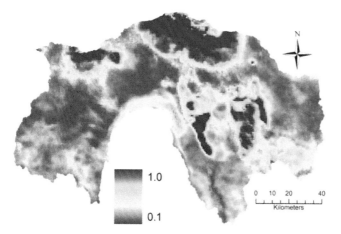

Land Use Type	Depletion fraction (*f*)
Water Bodies	0.87
Sparse Vegetation	0.27
Bushlands	0.29
Rainfed, Maize	0.30
Irrigation, Sugarcane	0.46
Irrigation; Bananas, coffee Mixed crops	0.60
Dense Forest	0.72

Fig. 5.2: Soil moisture depletion fraction (defined using average values of the dry month of January of 2008, 2009 and 2010) in the Upper Pangani River Basin for selected land use types.

Saturated zone

Apart from the net precipitation component $((1\text{-}c_r)\times P_e)$, the saturated zone receives water from the unsaturated zone when the soil moisture S_u reaches field capacity $(S_{u,max})$. Excess overflow (Q_u) is routed to the groundwater using a recession factor, K_u.

The saturated zone consists of three linear outlets which are separated by $S_{s,min}$ to represent the minimum storage level, $S_{s,q}$ to represent quickflow threshold and $S_{s,max}$ to represent rapid subsurface overflow. The flows are routed using K_o, K_q and K_s calibration coefficients respectively.

When the groundwater storage (S_s) exceeds the $S_{s,max}$, then saturation overland flow (Q_{of}) occurs:

$$Q_{of} = \max\left(S_s - S_{s,\max},0\right)/K_o \qquad (5.8)$$

where K_o is the overland flow recession constant.

The second groundwater flow component is the quick groundwater flow (Q_{qf}). It is assumed to be linearly dependent on the S_s and a quick flow threshold $S_{s,q}$ determined through calibration (Eq. 5.9).

$$Q_{qf} = \max\left(S_s - S_{s,q},0\right)/K_q \qquad (5.9)$$

where K_q is the quick flow recession constant.

The third component is the slow groundwater flow (Q_{sf}) which is dependent on the S_s levels

$$Q_{sf} = \left(S_s\right)/K_s \qquad (5.10)$$

where K_s is the slow flow recession constant.

K_o, K_q, K_s equal to 1, 2 and 28 respectively and were determined from recession curve analysis (where 1 unit is equal to the 8-day time step).

The maximum saturation storage $(S_{s,max})$ is a key variable that determines the dominant hydrological processes in the saturated zone. Three hydrological zones can be delineated from $S_{s,max}$, i.e. wetland, hill-slope and snow/ice zone. When $S_{s,max}$ is low, the saturation excess overland flow is dominant. This is characteristic for wetland systems described in detail by Savenije (2010). It occurs in the low lying areas of the Pangani river basin where slopes are modest, or with shallow groundwater levels. During a rainfall event, there is no adequate storage of groundwater leading to saturation excess overland flow. The wetland system is therefore dominated by Q_{of} and as such the $S_{s,max}$ is set very low or at zero (fully saturated areas) and c_r at 1.

As the elevation and slope increases, the groundwater depth as well as the $S_{s,max}$ increase gradually. This is characteristic of the hill-slope system where storage excess subsurface flow is the dominant runoff mechanism. Topographic indicators can be used to identify and separate this zone from the wetland system (where $S_{s,max}$ is near zero). Recently developed indices that can be used include the elevation above the nearest open water (H_o) (Savenije, 2010), or the Height Above the Nearest Drainage (HAND) (Nobre et al., 2011; Cuartas et al., 2012). The first topographic indicator, H_o (elevation above the nearest open water) is used in this study. H_o is derived from the level where groundwater storage is low or near zero. This was estimated from 92 groundwater observation levels located in the lower catchments of the river basin (Fig. 5.3).

Fig. 5.3 shows the delineation of the dominant hydrological processes in the Upper Pangani River Basin, including the wetland and hillslope (includes snowmelts at the peak of the mountains).

a)   b)

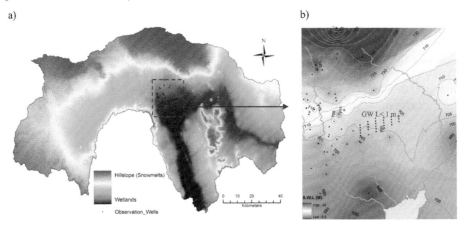

Fig. 5.3: (a) Wetland - Hillslope (Snowmelt) hydrological system (b) Shallow groundwater observation wells with mean surface water levels (0.3 - 40 m) in the lower catchments of the Upper Pangani River Basin for the period 2008-2010.

$S_{s,max}$ is not completely available for groundwater storage due to the soil texture (porosity and soil compression). According to Gerrits (2005), the maximum groundwater storage, $S_{s,max}$ [mm] for hillslope can be estimated using the natural log function of water storage depth, H_s (Eq. 5.11).

$$S_{s,\max} = 25 \times \ln H_s$$ (5.11)

where H_s [m] is the normalized DEM above H_o (where active groundwater storage is assumed zero). It is noteworthy that the wetland system can still exist along the drainage network of river system beyond H_o. This is possible since the H_s would still ensure a low groundwater storage ($S_{s,max}$) which makes the wetland system the dominant hydrological process. As observed in Fig. 5.3, the middle catchment forms the transition from the wetlands to the hillslope. It is noted that the hydrological landscape, plateau (dominated by deep percolation and hortonian overland flow) described in detail by Savenije (2010) is not existent on the slopes of Kilimanjaro and Meru, the higher elevations are forested and is active in the rainfall - runoff process. It is therefore modeled as forested hillslope.

The third zone delineated is the snowmelt. The amount of snow in the river basin is limited to the small portion of the mountain peaks of Mt. Kilimanjaro and Mt. Meru. The snowmelt occurs at elevation ranges of 4,070 m.a.s.l to 5,880 m.a.s.l and is derived from the land use map (Kiptala et al., 2013a). During rainfall seasons, the snow is formed and stored in the land surface. During the dry season, the snow melts gradually to the soil moisture and to the groundwater. This is unlike the temperate climate where a lot of snow cover is generated during the winter seasons which may result in heavy or excess overland discharge during the summer seasons. Furthermore, Mt. Kilimanjaro has lost most of its snow cover in the recent past due to climate variability/change, with significant snow visible only on the Kibo Peak (Misana et al., 2012). According to Grossmann (2008) the snowmelt contribution to groundwater recharge is insignificant in the Kilimanjaro aquifer. Simple representation of snowmelt can therefore be made using the hillslope parameters where the precipitation is stored in the unsaturated zone ($c_r = 1$ for snow) as excess unsaturated storage. The snowmelt is thereafter routed by K_u (unsaturated flow recession constant) to the groundwater over the season. This model conceptualization enables the hydrological model to maintain a limited number of parameters.

Interaction between the two zones

Capillary rise only occurs when groundwater storage is above a certain level, $S_{c,min}$. $S_{c,min}$ can be a fixed or a variable threshold value of the groundwater storage (S_s). Winsemius et al. (2006) adopted a fixed value of 25 mm as the $S_{c,min}$ for the Zambezi River basin. Since $S_{s,max}$ (from Eq. 5.11) is a function of H_s, a fixed threshold is not possible in this study. $S_{c,min}$ is made a function of groundwater storage S_s to provide a spatially variable threshold through calibration over the river basin. Capillary rise above this threshold is estimated on the basis of the balance between water use needs at the unsaturated zone and water availability in the saturated zone. Actual capillary rise is determined implicitly using the maximum capillary rise C_{max} (calibration parameter for each land use type), evaporation depletion ($E + T$) and the available groundwater storage S_s. Below $S_{c,min}$, a minimal capillary rise C_{min} is possible and is assumed to be zero for this study (timescale of 8-day is assumed low for substantial C_{min} to be realized).

$$C = \min\left(C_{max}, (E+T), S\right) \rightarrow if\left(S_s \geq S_{c,min}\right) \tag{5.12}$$

where the active groundwater storage for capillary rise, $S = S_s - S_{c,min}$.

However, since the capillary flow is low compared to water use for some land use types, supplementary blue water from river abstractions (Q_d) is required in the system. The third blue water storage term S_b, is introduced to regulate blue water availability from capillary rise, C, and river abstractions, Q_d. River abstractions include water demands from supplementary irrigation, wetlands and open water evaporation for lakes or rivers derived directly from the river systems.

$$Q_d = (Q_b - C) \rightarrow if\left(S_b \leq Q_b\right) \tag{5.13}$$

$$Q_d = 0 \rightarrow if\left(S_b > Q_b\right) \tag{5.14}$$

where Q_b is the blue water required to fill the evaporation gap that cannot be supplied from the soil storage. For irrigated croplands, Q_d is assumed to represent the net irrigation abstractions in the river basin. The assumption is based on the 8-day timestep that is considered sufficient for the return flows to get back to the river systems, i.e. the flow is at equilibrium. Q_d is therefore modeled as net water use in the river system.

Since river abstractions mainly occur in the middle to lower catchments and the accumulated flow would have a resultant net effect equivalent to the total simulated discharge, Q_s at a downstream outlet point or gauge station (Eq. 5.15 and 5.16).

$$Q_{s1} = Q_{of} + Q_{qf} + Q_{sf} \tag{5.15}$$

$$Q_s = Q_{s1} - Q_d \tag{5.16}$$

5.2.4   Sensitivity and uncertainty analysis

Since a number of assumptions were introduced to simulate the hydrological processes in the basin, a sensitivity analysis was performed to assess the influence of model input parameters to the variation of model performance. The parameter adjustments were done during the calibration process manually by trial and error. Some parameter values where manually altered within parameter ranges while others were calibrated freely. According to Lenhart et al. (2002), the parameter sensitivity can be achieved by varying one parameter at a time within the parameter range or using a fixed percentage change of the base value while holding the others fixed. Three parameter values; interception threshold (D), separation coefficient of net precipitation between the unsaturated and saturated zones (c_r) and the quick flow components (q_c) were varied within the parameter ranges. Three parameter values for maximum storage in the unsaturated zone $(S_{u, max})$, maximum storage in the saturated zone $(S_{s, max})$ and maximum potential capillary rise (C_{max}) that were calibrated freely were varied by a fixed change of the base value. The other three parameter values representing runoff timescales (K_o, K_q, K_s) were also varied by a fixed value from the estimates determined from the recession curve.

A sensitivity coefficient was computed to represent the change in the response variable that is caused by a unit change of an input variable, while holding the other parameters constant (Gu and Li, 2002). The sensitivity coefficient (SC) was normalized by reference values representing the range of each output and input variables to give the sensitivity index (SI) represented by Eq. (5.17).

$$SI = \left(\frac{y_i - y_0}{x_i - x_0} \right) \left(\frac{x_i + x_0}{y_i + y_0} \right) \tag{5.17}$$

where x_0 and y_0 are the base input parameter value and model output from the final model calibration respectively; x_i and y_i are the varied input parameter and the corresponding model output, respectively. SI makes it feasible to compare the results of different input parameters independent of the chosen variation range (Lenhart et al., 2002; Bastiaanssen et al., 2012). The SI can be positive or negative depending on the co-directional response of the model performance to the input parameter change. The absolute higher SI values indicate higher sensitivity.

5.2.5 Model performance

The modified STREAM model was calibrated and validated against measured daily discharge data from five gauging stations in the basin. One discharge gauge station, 1dd55, had a lot of missing data. Nevertheless, the limited information from this station, most upstream and the only one in the upper Mt. Meru, was useful in the calibration process of the downstream gauge stations. Additional downstream outlet points (dummy) were included for water balance analysis (See Fig. 5.9).

The daily discharge data were aggregated to 8-day time scale for the period 2008 - 2010. Since the secondary data from remote sensing (ET_a and f) were available for only 3 years, 1 year of data was used for calibration while the remainder of 2 years data used for the validation. An initial 1 year (46 simulations) was used as warm-up period to stabilize the model parameters using the mean input values. In total, the model was simulated for 184 time steps (4-year period). The following goodness to fit statistics were used to evaluate the model performance. The Nash-Sutcliffe model efficiency coefficient (E_{ns}) (Nash and Sutcliffe, 1970), Mean Absolute Error (MAE) and the Relative Mean Square Error (RMSE) in Eq. (5.18), Eq. (5.19) and Eq. (5.20) respectively.

$$E_{ns} = 1 - \frac{\sum_{i=1}^{n} (Q_s - Q_o)^2}{\sum_{t=1}^{n} (Q_o - \bar{Q}_o)^2} \tag{5.18}$$

where Q_s and Q_o are simulated discharge and observed discharge, $\bar{Q}_o$ is the mean of the observed discharge and n is the discharge data sets (n = 46 calibration; n = 92 validation).

$$MAE = \frac{1}{n} \sum_{i=1}^{n} |Q_s - Q_o| \tag{5.19}$$

$$RMSE = \sqrt{\frac{\sum_{i=1}^{n}(Q_s - Q_o)^2}{n}} \qquad (5.20)$$

Since the model priority objective is to simulate low flows, the E_{ns_ln} was also evaluated using natural logarithm of the variables in Eq. (5.18). The E_{ns} values range $[-\infty, 1]$, with 1 being the optimum (Ehret and Zehe, 2011). The range of MAE and RMSE is $[0, \infty]$, with zero being the optimum (Murphy, 1995). The model is optimized using these parameters to achieve a balance between the correlation, the bias, and the relative variability in the simulated and observed discharge (Gupta et al., 2009). The model estimates for irrigation water use $(Q_{b(I)})$, defined as Q_b for all the irrigation land use classes, were also compared with the field data on net irrigation water use from the river basin agency, Pangani Basin Water Office.

5.2.6 Scenario development

In Pangani River Basin, blue water use is currently over-exploited (Kiptala et al., 2013b). The implication for additional water allocation on stream flow to the nationally important hydropower stations needs to be known. This may also result in water savings or tradeoffs with other interventions or water uses. The crop yields for rainfed and supplementary irrigated lands are also low leading to low crop water productivity (Makurira et al., 2010). A few water management scenarios targeted on water savings and improved crop water productivity is explored using the modified STREAM model. They include i) Water saving through increased irrigation water efficiency, ii) increased crop productivity for rainfed lands, and iii) modifying the landscape for increased agricultural production.

To meet the first objective, the non-beneficial component of evaporation (soil evaporation) for irrigated landscapes is targeted for reduction. Soil evaporation (E_s) can account to up to 40% of evaporation depletion $(E+T)$ in irrigated landscapes (Bastiaanssen et al., 2012; Burt et al., 2001). In Pangani River Basin, located in the tropical climate, the irrigation system used by smallholder farmers that conveys water using small earthen furrow canals may have high levels of E_s. It is noteworthy that interception (I) also includes shallow (fast) soil evaporation that is implicitly derived only from precipitation. For demonstrative purposes, a reduction of 5% in $E + T$ for supplementary irrigated mixed crops is targeted (Scenario 1). The reduction represents about 15% of E_s if we assume a conservative E_s of 30% of $E + T$ in the supplementary irrigation systems. There are several methods for reducing E_s. They may include the lining of the main canals or using more efficient micro-irrigation systems. Further reduction can also be achieved by either straw or mechanical mulching (Prathapar and Qureshi, 1999; Zhang et al., 2003).

To meet the second objective, productive transpiration for rainfed maize (highland) is increased by 30% (Scenario 2a). According to Makurira et al. (2010), the crop water productivity for smallholder rainfed farms can be improved by using systems innovations (SIs). The study was done in Makanya catchments within the Pangani River Basin. The SIs used combined runoff harvesting with in-field trenches and soil bunds which resulted in an increase of transpiration of 47%. The SIs aimed also at preventing soil and nutrient loss. An increase in T would result in an increase in biomass production and thus crop yields (Steduto et al., 2009). The rainfed maize in the highland areas was targeted due to the relative high precipitation during the rainy seasons. In-field trenches and soil bunds (*Fanya juus*) is normally associated with high infiltration levels and higher soil moisture retention (Kosgei et al., 2007; Makurira et al., 2010). An additional increase in $S_{u,max}$ of 30% is also investigated in addition to the increased transpiration for highland rainfed maize and coffee (Scenario 2b).

For the third objective, the area for irrigated sugarcane is doubled to its potential (Scenario 3). Currently, TPC irrigation scheme covers an area of 8,000 ha, for which 7,400 ha is under sugarcane cultivation with the reminder providing the irrigation services. The potential irrigation area is estimated at 16,000 ha constrained by limited water resources. The expansion of the irrigation system is of great interest in the basin due to the high sugar demand and increasing potential for bio-fuels.

5.3 RESULTS AND DISCUSSION

5.3.1 Calibration and validation results

Figs. 5.4 & 5.5 show the comparison of the observed and simulated hydrographs and the average precipitation for five outlets (gauge stations) in the Upper Pangani River Basin. The figures provide a visual inspection of the goodness of fit of the data with an additional scatter plot for the most downstream outlet (1dd1). The model simulates the base flows very well both during the calibration and validation periods. The peak flows for the larger streams (1dd54, 1dd1) were better simulated than for the smaller streams (1dc8a, 1dc5b, 1dc11a). It is to be noted that the observed discharge data is also subject to uncertainty which is more pronounced for the smaller streams. The remotely sensed data, ET_a and f also have a higher uncertainty during the rainy season (peak flow season). This is the period when most clouded satellite images exist and the cloud removal process is subject to uncertainty (Kiptala et al., 2013b).

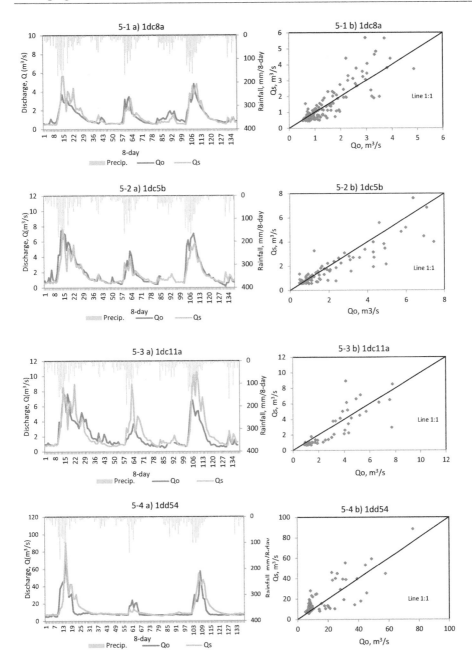

Fig. 5.4: (1 - 4) (a) Comparison of observed (Qo) and the simulated discharge (Qs) and precipitation at the outlet points for calibration period 2008 (8-day periods 1 - 46) and validation 2009, 2010 (8-day periods 47 - 138) in the Upper Pangani River Basin; and (b) the corresponding scatter plots of Qo and Qs for four upstream gauge stations.

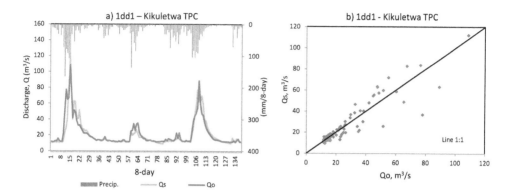

Fig. 5.5: (a) Comparison of observed (Qo) and simulated discharge (Qs) and precipitation for calibration period 2008 (8 day periods 1–46) and validation 2009, 2010 (8 day periods 47–138) in the Upper Pangani River Basin; and (b) the corresponding scatter plot of Qo and Qs for the most downstream gauge station.

Table 5.1 shows the performance model results for the validation and calibration of the modified STREAM model in the Upper Pangani River Basin. The Nash-Sutcliffe Efficiency, E_{ns} for the calibration period scored > 0.5 (except for 1dd11a, where $E_{ns} = 0.46$) which is indicative of good model performance. In the validation period, two outlet points had scores < 0.5 (1dd11a - 0.33 and 1dd54 - 0.42) which indicates a moderate performance. The Nash-Sutcliffe Efficiency for natural logarithm, E_{ns_ln}, which emphasizes the base flow, resulted in better results with all outlet points scoring more than 0.6. There was a slight reduction in E_{ns_ln} in outlet points 1dd54 (calibration) and 1dd8a, 1d5b (validation) but overall the model performance on the low flows was good.

Table 5.1: Model performance for the modified STREAM model for Upper Pangani River Basin.

Station	Calibration				Validation			
	E_{ns}	E_{ns_ln}	MAE ($m^3 s^{-1}$)	RMSE ($m^3 s^{-1}$)	E_{ns}	E_{ns_ln}	MAE ($m^3 s^{-1}$)	RMSE ($m^3 s^{-1}$)
1dc8a	0.63	0.68	0.73	0.92	0.72	0.68	0.62	0.36
1d5b	0.75	0.77	0.74	1.09	0.81	0.78	0.57	0.23
1dd11a	0.46	0.64	0.84	1.14	0.33	0.69	0.94	0.88
1dd54	0.70	0.60	2.31	8.06	0.42	0.61	1.99	5.84
1dd1	0.84	0.90	2.08	9.34	0.83	0.90	1.74	4.78

MAE ranged between 0.62 $m^3 s^{-1}$ and 2.08 $m^3 s^{-1}$ for the larger streams in the calibration period. A big difference is observed between the RMSE and MAE (up to four times) for the downstream stations 1dd54 and 1dd1 during the calibration period. The result is indicative of large (noisy) variations between the simulated and observed discharges. Fig. 5.4 also shows that the large deviations arise during the rainy periods (Masika and Vuli seasons). This may be attributed to the uncertainties of the remote sensing data in the clouded periods (rainy days). Such uncertainties can be avoided by using passive microwave imagery (Bastiaanssen et al., 2012). Furthermore, the river gauging stations are poorly maintained in the river basin. The discharge rating curves are also not regularly updated despite the changes in the river regime. Model conceptualization assumptions such as irrigation water use and return flows may also not coincide in space and time with the actual processes in the river basin. Errors in boundary conditions on the representation of groundwater may also occur if they do not coincide with the river systems.

5.3.2   Sensitivity analysis

The sensitivity analysis of the input parameters is given in Table 5.2. The sensitivity index (SI) was analyzed using the RMSE and MAE model performance indicators for the entire simulation period using the discharge measurements at outlet point (1dd1). The base input values (x_0) were the final calibrated values that were varied by a fixed or percentage change $(x_1$ or $x_2)$. Decrease in $S_{u,max}$ by 25% resulted in the highest SI of -1.97 for RMSE. However, a similar increase of 25% did not have any significant change in model output. The sensitivity is mainly attributed to the overland flow that is influenced by the water storage in the unsaturated zone. Similar changes in $S_{s,max}$ also resulted in moderately high sensitivity for both RMSE and MAE. This is mainly because the saturated zone controls all the runoff components. Separation coefficient c_r that separates the net precipitation between unsaturated and saturated zones and the quick flow coefficient, q_c had high sensitivity. The values used $c_r = 0.75$ and $q_c = 0.75$ (aggregated averages) for various land use types were generally derived from previous modelling experiences and where based on the soil type and land use.

The soil moisture depletion fractions (f) were derived from the SEBAL model for various land use types. An aggregated average f value of 0.33 was adopted from the mean values for the land use types that ranged between 0.2 for natural land use types to over 0.6 for irrigated agriculture (also see Fig. 5.2). These parameters resulted in minimum sensitivity since the ranges used ($\pm25\%$ of the base values) where reasonable within the derived estimates from remote sensing. The runoff timescales parameters K_o and K_q also had low sensitivity because the flow times were short and within the estimates derived from the recession curves. The timescale K_s for slow groundwater flow that has a higher flow times had a moderate sensitivity. A lower timescale for K_o of 1 time step (8-days) may introduce some uncertainty if the model was used to simulate flood events that are critical at shorter timescales of 1 - 2 days. However, for hydrological processes that characterize agricultural water use such as irrigation scheduling or dry river flows, the uncertainty is minimal.

The maximum capillary rise (C_{max}) was calibrated through a water balance process to maintain the evaporation depletion $(E + T)$. An aggregated average value of 2 mm day^{-1} was achieved and ranged between 1.1 mm day^{-1} for woodland landscape in semi-arid areas to a maximum of 2.8 mm day^{-1} in the natural dense forest in humid climate. The calibrated values were within the ranges for natural vegetation reported in literature (Shah et al., 2011). In natural and rainfed systems, only C_{max} was calibrated to maintain the evaporative capacity of the unsaturated zone. The actual capillary rise (C) would not change with an increase in C_{max}. However, a decrease in C_{max} would constrain C, thus resulting in lower soil moisture conditions in the unsaturated zone. For irrigated land use types, the evaporative capacity $(E + T)$ is maintained by both C and irrigation (Q_d). The changes in C due to high or lower C_{max} threshold will correspond to a similar change in Q_d. C_{max} was therefore a less influential parameter with low sensitivity in natural vegetation. Interception threshold, D showed also low sensitivity to changes within the parameter range. D was computed on a daily basis using the interception threshold for various landuse types derived from literature. However, the actual interception is more dependent on the daily variability of rainfall than the total interception threshold. Similar findings were observed by De Groen and Savenije (2006). While the interception threshold is not an influential parameter, actual interception (I) is still important water balance component as the water for the other processes is dependent on the net precipitation after interception (Makurira et al., 2010).

Table 5.2: Sensitivity of model performance due to change in model input parameters.

Parameter	Input Values			Resulted RMSE (m³ s⁻¹)					Resulted MAE (m³ s⁻¹)				
	x_1	x_0	x_2	y_1	y_0	y_2	SI (x_1)	SI (x_2)	y_1	y_0	y_2	SI (x_1)	SI (x_2)
D [mm/day]	0	2.5	4	8.8	6.9	7.1	-0.12	0.02	2.0	1.8	1.8	-0.12	0.01
$S_{u, Max}$ [mm]	262	350	438	12.4	6.9	6.9	*-1.97*	0.04	2.1	1.8	1.8	-0.47	0.19
$S_{s, Max}$ [mm]	150	200	250	9.3	6.9	8.0	*-1.00*	0.48	2.2	1.8	2.2	-0.64	0.66
c_r [-]	0	0.75	1.0	202.5	6.9	16.6	*-0.93*	*1.25*	9.6	1.8	2.9	-0.69	0.71
q_c [-]	0	0.75	1.0	39.7	6.9	7.7	-0.70	0.20	5.4	1.8	1.9	-0.50	0.07
C_{max} [mm/day]	1.5	2.0	2.5	7.2	6.9	7.1	-0.14	0.08	2.0	1.8	1.8	-0.34	0.00
f [-]	0.25	0.33	0.41	6.9	6.9	7.1	0.00	0.07	1.8	1.8	1.8	0.00	0.10
K_o [8-day]	-	1	2	-	6.9	6.9	-	0.00	-	1.8	1.8	-	0.02
K_q [8-day]	1	2	3	7.0	6.9	7.0	0.00	0.02	1.8	1.8	1.9	-0.07	0.08
K_s [8-day]	20	28	35	7.4	6.9	7.5	-0.19	0.27	2.2	1.8	2.0	-0.49	0.27

SI in italics denotes high sensitivity

5.3.3 Model interpretation

Interception and Evaporation depletion

There is general consensus that actual interception (I) is a key component in hydrology and water management. I influence the net precipitation and therefore the amount of water available for evaporation $(E+T)$. Evaporation depletion $(E+T)$ influences the stream flow dynamics and is the manageable component of ET_a in biomass production. Therefore, there is a need to distinguish $E+T$ from the calculated I as a deficit of total ET_a (SEBAL), Fig. 5.6.

The mean annual I ranged between 8 - 24% of the total evapotranspiration. The land use types in the upper catchments, e.g. forest, rainfed coffee and bananas, had higher I. Irrigated sugarcane and natural shrublands located in the lower catchments had lower I. The variation is mainly influenced by the maximum threshold (D) and the rainfall (intensity and frequency) which are relatively higher for land use types in the upper catchments. The forest interception average estimate of 24% of the total evapotranspiration (or 22% of the total rainfall) is comparable with field measurements from previous studies that found forest canopy interception of about 25% of the total rainfall in a savannah ecosystem in Africa (Tsiko et al., 2012).

Q_b contributions, e.g. irrigation, also enhanced the evaporation depletion $(E+T)$ component of ET_a resulting in relatively lower I for irrigated croplands. Any intervention to change I would influence antecedent soil moisture conditions especially during small rainfall events (Zhang and Savenije, 2005). This may influence the productivity of $E+T$ and/or the stream flow generation in the river basin. However, more research is required to estimate explicitly the changes in I from certain field based interventions. The outcome of such studies maybe incorporated in the STREAM model.

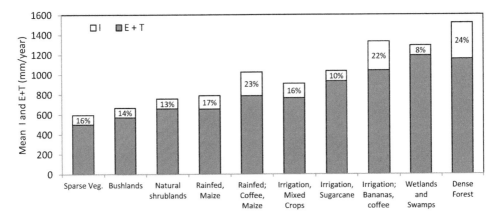

Fig. 5.6: Mean Interception, I, and evaporation depletion, $E + T$ for different land use classes in Upper Pangani River basin for Period 2008 - 2010.

Blue and green water use

Figs. 5.7 and 5.8 shows the resultant blue water use (Q_b) and the direct contribution of precipitation (Q_g) to the ET_a (actual evapotranspiration) for various land use types. Q_b is closely related to the land use and the ET_a as observed in Figs. 3.5 & 4.2. Water bodies (lakes and reservoir) and the wetlands have the highest Q_b, contributed by the high open water evaporation. The average Q_b for water bodies is approx. 56% of the ET_a with a maximum of 74% (1,642 mm yr^{-1}) observed at the lower end of the NyM reservoir. The Q_b is high in the NyM reservoir because of the high potential evaporation attributed to hotter climatic conditions and lower precipitation levels in the lower catchments. Wetlands and swamps located in the lower catchments also resulted in high Q_b of approximately 42% of ET_a. In irrigated croplands, the Q_b was also moderately high with a range of between 20% for irrigated mixed crops and bananas in the upper catchments, and 44% for irrigated sugarcane in the lower catchment.

Rainfed crops and natural vegetation including the forests had a lower Q_b, mainly stemming from groundwater (and snow melts). Sparse vegetation, bushlands, grasslands, natural shrublands had Q_b contributions of less than 1% of total ET_a, while rainfed maize (middle catchments) and rainfed coffee (upper catchments) had Q_b contributions of 2% and 7% of ET_a respectively. Dense forest and Afro-Alpine forest had slightly higher Q_b contributions (ranging between 7 - 9 %) attributed mainly to the availability of groundwater from snow melts in the upper mountains.

Notable higher Q_b was experienced in the dry year of 2009 (as shown by the error bars in Fig. 5.8). This is attributed to higher potential evaporation from relatively drier weather conditions. The lower precipitation during this period also resulted in increased groundwater use for the afro-alpine and dense forest land uses in the upper catchments. For instance the Q_b contribution to ET_a for dense forest increased from 5% in 2008 (a relatively wet year) to 10% in 2009. The enhanced Q_b for the irrigated croplands during 2009 is also attributable to the higher potential evaporation and limited precipitation that increased the irrigation water requirement. This is illustrated by irrigated sugarcane where Q_b increased from 35% in 2008 to 55% in 2009. Q_b for supplementary irrigation also increased from 14% to 29% during the dry year. The Q_b for year 2010 was in general average for all land use types which is indicative of the average weather conditions that prevailed during the year.

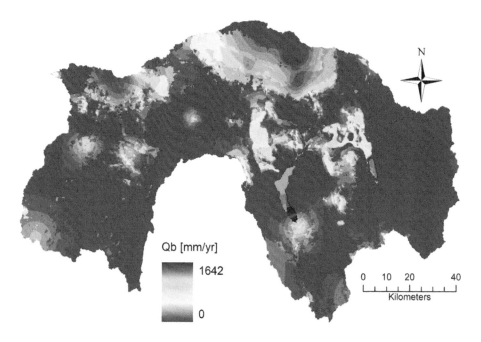

Fig. 5.7: Spatial Variability of blue water use (Q_b) averaged over 2008 - 2010 in the Upper Pangani River Basin.

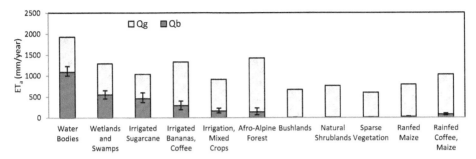

Fig. 5.8: ET_a **and the corresponding** Q_g **and Qb water use for selected land use types averaged per year over 2008 - 2010 in the Upper Pangani River Basin (Error bar indicates the upper and lower bounds for mean** Q_b **for dry year 2009 and wet year 2008 respectively).**

Irrigation water use

This section presents the model results for supplementary irrigation water use ($Q_{b(I)}$) as estimated at various outlet points (gauging stations) in the river basin. The annual irrigation abstractions, predominant during dry seasons, were accumulated and the average mean for the period 2008-2010 is presented in Fig. 5.9. Six gauge stations and three additional points (accumulation points for Kikuletwa, Ruvu and Lake Jipe) were also considered. The annual net irrigation (in million cubic meters) was converted to m^3 s^{-1} to allow easier comparison with the discharge data in section 5.3.1.

The $Q_{b(I)}$ ranges from 0.06 m^3 s^{-1} on the smaller streams to a total of 3.4 m^3 s^{-1} and 4.2 m^3 s^{-1} in the outlets of the Ruvu and Kikuletwa river systems respectively. A significant irrigation abstraction of 1.5 m^3 s^{-1} was observed by the TPC sugarcane irrigation system, the largest single irrigation scheme in the river basin. The total $Q_{b(I)}$ upstream of NyM reservoir was estimated at 7.6 m^3 s^{-1}, which represents approximately 50% of the low flows in the Upper Pangani River Basin.

Open canal irrigation is the commonly used irrigation technique in the Upper Pangani River Basin. There are an estimated 2,000 small-scale traditional furrow systems, some 200 - 300 years old (Komakech et al., 2012). According to records at the Pangani Basin Water Office, approximately 1,200 of these abstractions have formal water rights. PWBO estimates that the total gross irrigation abstraction is approximately 40 m^3 s^{-1}. The irrigation efficiencies of the irrigation systems range between 12 - 15% (Zonal Irrigation office, Moshi). Here, we adopted higher irrigation efficiency limit of 15% to compensate for any uncertainties that may arise from the higher irrigation efficiencies in the larger irrigation schemes. The field estimates provides net irrigation consumptions of approximately 6 m^3 s^{-1} (using 15% efficiency) and about 79% of the $Q_{b(I)}$ model estimates (19% efficiency). The water leaks in the traditional furrow canals flows back to the river system. The capacity and ability of the river basin authority to monitor actual water abstraction is limited. However, considering these uncertainties, the modeled net irrigation abstraction was reasonably close to field net irrigation estimates for the Upper Pangani Basin.

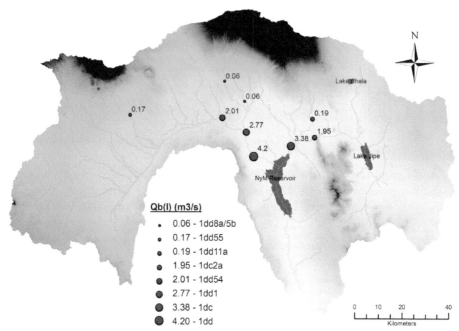

Fig. 5.9: Total net irrigation water use $(Q_{b(I)})$ estimated upstream of the gauge stations using modified STREAM model in the Upper Pangani River Basin (averaged over 2008-2010).

Open water evaporation

The blue water use by the water bodies $(Q_{b(w)})$ upstream of NyM reservoir was also estimated using the modified STREAM model. $Q_{b(w)}$ is the net open water evaporation from blue water which would otherwise flow into the NyM reservoir. The water bodies considered include wetlands (98 km^2), Lake Jipe (25 km^2) and Lake Chala (4 km^2). The total mean $Q_{b(w)}$ were estimated to be 53.6×10^6 m^3 yr^{-1} (1.7 m^3 s^{-1}) and 22.1×10^6 m^3 yr^{-1} (0.7 m^3 s^{-1}) in the Ruvu and Kikuletwa river systems, respectively. The total $Q_{b(w)}$ (12% of low flows) may also provide an insight into ecosystem services or benefits provided by the natural water bodies compared with the alternative uses, such as irrigation or hydropower in the downstream part of the river basin.

5.3.4 Future water management scenario using modified STREAM model

The previous sections illustrate how the modified STREAM model provides spatial information on the water use (green and blue) under current situation. The information is useful especially in monitoring unregulated irrigation water use. The model also provided useful information on the implication of future water use management scenarios in the river basin. Table 5.3 shows the real impact of the interventions on the water resources under the scenarios defined in section 5.2.6. The changes in surface runoff were evaluated from the outlet points (1dc & 1dd) upstream of NyM reservoir, Upper Pangani River Basin (Fig. 5.9).

Table 5.3: Impact of three water management scenarios on the surface runoff.

Scenario		Action	Impact on outflow (Mm^3 yr^{-1})		
			Total	Base flow	Overland flow
1	Reduce E_s	Reduce E_s for supplementary irrigation (mixed crops) by 15% or approximately 5% of transpiration	37.8	34.5	3.2
2 (a)	Increase T	Increase T by 30% for rainfed maize in the highlands areas	-84.2	-77.6	-6.6
(b)		plus 30% increase in $S_{u,max}$	-87.0	-76.9	-10.1
3	Modify area	Double sugarcane irrigated area (additional 7,400ha)	-53.9	-53.3	-0.6

If soil evaporation is reduced in irrigation systems (Scenario 1), real water saving of 37.8×10^6 m^3 yr^{-1} can be achieved. The additional water saved (4% of total river flow), mostly groundwater flow can be utilized in the expansion of the irrigated sugarcane (scenario 3). Scenario 1, alternatively, could also release additional base flow that may be required for other water uses that include environmental and/or downstream hydropower demands. Financing of the required interventions can also form a basis for basin-wide trade-off negotiations between downstream and upstream water users.

Scenario 2(a) investigates the implications of up scaling system innovations (SIs) for the rainfed maize cultivated in the highlands. In the area targeted, mixed farming of maize and coffee is practiced, and covers an area of 72,300 ha (Kiptala et al., 2013a). Half of this area is under maize cultivation. This intervention would result in additional water use of 84 $\times10^6$ m^3 yr^{-1}, which is about 10% of the total river flows. The model simulation shows that the water use will be derived from base flow. However, small-scale runoff harvesting devices can be used to store overland flow to supplement blue water needs during the dry seasons. Scenario 2(b) shows that an increase in both T and $S_{u,max}$ would result in slightly higher overland flow water use. This will not only increase water availability in the unsaturated zone for transpiration, but also reduce the soil and nutrient losses normally associated with higher overland flows.

In scenario 3, the increase in the sugarcane irrigated area by 7,400 ha required an additional 53.9 Mm^3 yr^{-1} in average of base flow. The volume required for each year: 45.6 Mm^3 yr^{-1} (2008), 68.6 Mm^3 yr^{-1} (2009) and 47.4 Mm^3 yr^{-1} (2010) varied with the climate conditions. This is about 4%, 11% and 6% of the total river flows in 2008, 2009 and 2010 respectively. An additional conveyance and drainage losses may increase the net water use. It was also observed that the total additional blue water required in scenario 3 is consistent with the current irrigation water use (Q_b) by the existing irrigation system.

5.4 Conclusions

This chapter presents a novel method of developing a spatially distributed hydrological model using blue and green water use at pixel scale. The methodology allows for unprecedented insights into hydrological processes at smaller scales of land use classes. The hydrological model was developed for a heterogeneous, highly utilized and data scarce landscape with a sub-humid and arid tropical climate. The blue water use was quantified by employing a time series of remotely sensed evapotranspiration data as input in STREAM model. The model was also constrained by satellite-based soil moisture estimates that provided spatially (and temporally) realistic depletion levels during the dry season. To further enhance model parameter identification and calibration, three hydrological landscapes wetlands, hill-slope and snowmelt, were identified using topographic data and field observations. Unrealistic parameter estimates, found for example in natural vegetation either through overestimation of satellite-based data or model structure, were corrected in the model conceptualization through the water balance (at pixel scale). The modified STREAM model provided a considerably good representation of supplementary blue water use, which is dominant in the Upper Pangani River Basin.

The model performed well on discharge, especially in the simulation of low flows. The Nash-Sutcliffe coefficient (E_{ns_ln}) ranged between 0.6 to 0.9 for all outlet points in both calibration and validation periods. At the outlet, the model performance was best $(E_{ns\_ln} = 0.90)$. The large difference between MAE and RMSE was indicative of large errors or noisy fluctuations (see Figs. 5.4 & 5.5) between actual and simulated discharges during the rainy seasons. This was mainly attributed to the uncertainties of the remote sensing data during clouded periods. The uncertainties may also have been exacerbated by possible errors in the hydro-meteorological data and model conceptualization. Model parameters that were freely calibrated for different land use such as maximum unsaturated and saturated storages $(S_{u,max}, S_{s,max})$, separation coefficient (c_r) and quick flow coefficient (q_c) resulted in high sensitivity. The model calibration of these parameters can be improved in future by field measurement or by analytical relationships.

The simulated net irrigation abstractions were estimated at 7.6 m^3 s^{-1} which represents approximately 50% of low flows. Model results compared reasonably well with field estimates with less than 20% difference. In addition, the model yields spatially distributed data on net blue water use that provides insights into water use patterns for different landscapes under different climate conditions. Blue water use (Q_b) contribution to ET_a during a dry year (2009) increased from 5% to 10% for dense forest, 35% to 50% for the wetlands and irrigated sugarcane, and 14% to 28% for supplementary irrigation compared to the wet year (2008). Three water management scenarios on water saving and increased crop productivity were also explored using the STREAM model. Reduced soil evaporation of 15% on supplementary irrigation system would result in real blue water savings of 37.8×10^6 m^3 yr^{-1} (4% of total river flows). The water saving could alternatively be used to expand the existing sugarcane irrigation scheme (7,400 ha on sugarcane) that required 6% of total river flows if its area is doubled. Up-scaling of systems innovation for highland rainfed crops to achieve a 30% increase in productive T

resulted in additional blue water requirement of 84×10^6 m^3 yr^{-1}. The additional water requirement can be generated from runoff harvesting and storage to save on the already over-exploited blue water resources. This information may form a basis for socio-economic trade-off analysis on the basis of which various basin strategies and financial mechanisms can be formulated for efficient, equitable and sustainable water resources management at the river basin.

The development of advanced methods of generating more accurate remotely sensed data should go hand in hand with ways to improve distributed hydrological models. Such methods may include the use of passive microwave imagery to generate cloud free ET_a estimates (Bastiaanssen et al., 2012). Future modelling improvements should also aim at simulating the model for longer time series using long term rainfall and RS data (evapotranspiration and soil moisture). The data could be based on stochastic or probabilistic techniques (Salas et al., 2003). In so doing, data can be interpreted in a way that is useful for management and decision-making.

Chapter 6

WATER PRODUCTIVITY[4]

Scarcity of information on water productivity for different water, land and ecosystems in Africa hampers optimal allocation of the limited water resources. This chapter presents an innovative method to quantify the spatial variability of biomass production, crop yield, consumptive water use and economic water productivity in a data scarce landscape of the Pangani river basin. For the first time, gross return from carbon credits and other ecosystem services are considered in the concept of Economic Water Productivity. Carbon credits and water yields provide insights into the water value society attaches to a certain cultural and/or natural land use activity. The analysis relied on open-access multi-temporal Moderate - resolution Imaging Spectroradiometer (MODIS) satellite data of 250-m and 8-day resolutions. Instead of using default MODIS products, actual evapotranspiration and biomass production were computed with the Surface Energy Balance Algorithm for Land (SEBAL) utilizing Monteith's framework for ecological production. Grid biomass production (kg ha^{-1}) was estimated and converted into crop yield and amount of carbon sequestered. Gross returns were estimated using conversion factors for crop yield, carbon assimilates and market prices.

The Economic Water Productivity for 15 land use types, including natural land covers providing ecosystems services, were thus computed. In agriculture, irrigated sugarcane and rice achieved the highest water productivities in both biophysical and economic values - well within the ranges reported in the literature. Rainfed and supplementary irrigated banana and maize productivities were significantly lower than potential (maximum) values, reflecting a wide spatial variability. As expected, in natural land cover, dense forests and wetlands showed the highest biomass productivities. Spatially explicit information of water productivity using both biophysical and economic indicators has the potential to provide a coherent and holistic outlook of the socio-environmental and economic values of consumptive water use in river basins, can identify areas where these values can be improved, and when coupled to a hydrological model, may be a basis for trade off analysis.

---

[4] This chapter is based on: Kiptala, J.K., Mohamed, Y.A., Mul, M.L., Bastiaanssen, W.G.M., Van der Zaag, P., 2016a. Mapping Mapping ecological production and gross returns from water consumed in agricultural and natural landscapes. A case study of the Pangani river basin, Tanzania. Submitted to *Water Research and Economics.*

6.1 INTRODUCTION

The competition over water resources is escalating in many river basins worldwide. Population growth and increasing food demands to meet not only local but also global needs, imposes high pressures on the world's fresh water resources. More water is required to provide for the increasing energy demands from hydropower and biofuels (e.g. De Fraiture et al., 2008). Competition over water is not limited to the agricultural, domestic and industrial sectors but also includes the natural environment. Ecosystem services provide essential functions that can be grouped in provisioning, regulating, as well as habitat and cultural services (e.g. Millennium Ecosystem Assessment, 2005; De Groot et al., 2012). While environmental water uses are fundamental for sustainable economic and social development in river basins, it is becoming clear that the natural environment consumes large amounts of the water resources, and that measures are required to safeguard natural capital (Rockström et al., 1999; Wackernagel et al., 1999; Monfreda et al., 2004).

The main water flux in a river basin is evapotranspiration, with a large part being consumed by the natural landscape. For example, in the Awash basin, Ethiopia, 49% of all evapotranspiration occurs in mosaic forested shrubland/grassland and closed to open shrubland (Karimi et al., 2015). An analyses on the water consumption of different land use classes for the Nile basin showed that savannah is the largest water consumer 38%, followed by pastures (9%), wetlands (7%) and rivers and lakes (7%) of the total evapotranspiration (Bastiaanssen et al., 2014). The total evaporative use of the Nile is thus for 61% explained by environmental systems. Hence, the natural environment consumes large portions of rainfall (i.e. green water), and water from inundations and shallow groundwater tables (i.e. blue water). It is therefore legitimate, if not imperative, to include the economic value of natural ecosystems in the economic water productivity of river basins. So far, most authors only considered agricultural goods when employing the concept of Economic Water Productivity (e.g. Barker et al., 2003; Rodriguez-Ferrero, 2003; Ali et al., 2007; Hellegers et al., 2009). Valuing water consumption in a river basin should go beyond marketable crops, and encapsulate financial rewards for the sequestration of atmospheric carbon, heat and other forms of ecosystem services. Payment for Ecosystem Services (PES) is an emerging topic that is gradually implemented by policy makers in several basins, both in developed and developing countries. Some PES programs involve contracts between consumers of ecosystem services and the suppliers of these services. However, the majority of the PES programs are funded by governments and involve intermediaries, such as non-government organisations (FAO, 2011).

Natural environments such as forests and wetlands provide a wealth of ecosystems goods and services in terms of delayed peak runoff due to retention of excess rainfall, sufficient catchment water yield that ensures base flow in the dry season, sustaining rainfall by means of atmospheric moisture recycling, reduced erosion and improved water quality, from which downstream societies benefit. Natural ecosystems good and services also provide a buffer to local communities during periods of droughts in semi-arid African landscapes (Enfors and Gordon, 2008). Degradation of such environments reduces such benefits significantly. Similarly, increased withdrawals in upper catchments directly affect downstream ecosystems, such as wetlands, riparian vegetation and delta and estuarine ecosystems. Because of undesirable upstream developments, most perennial tributaries in our study area, i.e. the Pangani River Basin in Tanzania, have actually become seasonal in the last few decades (IUCN, 2009).

Costanza et al. (2014) estimates the global value of ecosystem services in 2011 has US$ 125 trillion a year, with a loss of up-to US$ 20 trillion per year from 1997 due to land use change. The study highlighted the magnitude of the ecosystem services and the urgent need for policies that promotes environmental conservation. In the United States, farmers are paid an amount of US$ 1.8 billion per year to conserve soil and water on 1.4 million hectare of "environmentally friendly" land, i.e. 1,285 US$ ha^{-1} yr^{-1} (Stubbs, 2014). Similarly, the Chinese government has a program to not clear forests, and compensates an amount of US$ 50 billion annually to local farming communities (Saah and Troy, 2015).

It is clear that optimal water use is indispensable in river basins that are closing with prevailing competition over water resources (Keller et al., 1998). Ecological and hydrological integrity is central to sustainable water resource use. Crop water productivity expresses the returns per unit of water consumed in river basins, and this can either be expressed biophysically, economically or socially (Igbadun et al., 2006; Zwart and Bastiaanssen, 2007; Yokwe, 2009; Molden et al., 2010; Van der Zaag, 2010; Bossio et al., 2011; Fereres et al., 2014). Therefore, water productivity (kg m^{-3} and $ m^{-3}) can be a key indicator to assess the effective use of water, although never as a sole purpose (Molden and Sakthivadivel, 1999).

Increasing the water productivity of crops is needed to achieve the broad objectives of increasing food production, income and livelihoods while maintaining ecological integrity – more crop per drop (Molden et al., 2010). Farm practices to increase physical water productivity are well documented in the literature. These include water harvesting, supplementary irrigation, soil water conservation, deficit irrigation and right use of fertilizers and pesticides amongst other interventions. Water productivity gains in agriculture would secure water resources for other landscape uses and ecosystem services (Kijne et al., 2009). Molden et al. (2010) identifies four priority areas where substantial increases of water productivity can be achieved at relatively low environmental and social costs per unit of water consumed. These include (i) low water productivity in areas with high poverty, (ii) areas of physical water scarcity where competition for water is high, (iii) areas with little water resource development where high returns can be achieved with additional water use, and (iv) areas of water-driven ecosystem degradation.

Environmental biomass production that has high environmental and social values often has little or no market price (Costanza et al., 1997; Hermans et al., 2006; Enfors and Gordon, 2007; Batjes, 2012; De Groot et al., 2012). The situation is poised to change due to the emerging market of carbon credits. The Kyoto Protocol allows the emission of carbon by industries into the atmosphere to be offset by sequestration in forests and other forms of permanent green landscapes (Sedjo, 2001). Carbon trading appeared as a potential business in the early 21st century, which more recently was hampered by the economic crisis since 2008 and fewer industrial activities resulting in a dramatic drop in carbon prices. Many economists, however, maintain that putting a price on carbon and allowing for carbon trading is a crucial element of the global climate policy (Pizer, 2002; Stern, 2007).

There is, however, a systematic lack of data on the water productivity of natural vegetation. Moreover, natural vegetation is highly heterogeneous with large spatial coverage. A Measurement-Reporting-Planning-Monitoring (MRPM) system is required to quantifying such water productivities. This chapter demonstrates that estimates based on remote sensing can achieve this. Much progress has been made recently using remotely sensed data in spatial water productivity analysis (Zwart and Bastiaanssen, 2007). In few cases these have been complemented by modelling approaches to estimate water productivity, Van Dam et al. (2006) combined a soil-physical simulation model with SEBAL and Mainuddin and Kirby (2009) used FAO's CROPWAT (Allen et al., 1998). At a global scale, water productivity models have been developed, WATPRO is a model developed for wheat water productivity (Zwart et al., 2010) and GEPIC (GIS – based Environmental Policy Integrated Climate model) is a agro-ecosystem model to assess global water productivity (Lui et al., 2009). Finally, remote sensing data on water productivity has been complemented by using crop yield statistics (McVicar et al., 2002). More recently, Yan and Wu (2014) and Zhang et al. (2015) presented the spatial-temporal crop water productivity for winter wheat in China using remotely sensed estimates of ET. Only one study reported on the economic water productivity using a combination of remote sensing and economic analysis for two main market crops, bananas and sugarcane (Hellegers et al., 2009). So far few studies included valuation and none so far have included natural land use types or non-market based crops.

Therefore, the objective of this chapter is to present a basin-wide analysis of the bio-physical (biomass and yield) and economic water productivity of both agricultural and natural land uses in a river basin. The methodology integrates actual field data and auxiliary crop information from the literature with remotely sensed data. The water productivity analysis is applied to a heterogeneous African landscape of the Pangani river basin in East Africa. The uncertainty of biomass production was also assessed. The explicit water productivity maps are presented using both non-market (biophysical) and market (economic) methods. This information can inform strategies for increasing the economic water productivity of certain land uses without impacting significantly on the environmental and social benefits. This approach will broaden the scope of interventions to improve the water productivity in agriculture while sustaining the ecosystem services and green growth in natural land systems.

6.2 MATERIALS AND METHODS

6.2.1 Actual evapotranspiration

The actual evapotranspiration (ET) data into the water productivity analysis were available at 250-m and 8-day resolutions for the period 2008, 2009 and 2010 for the Upper Pangani River Basin. The ET data were computed in Chapter 4 based on the SEBAL algorithm (Bastiaanssen et al., 1998).

6.2.2 Biomass production

Biomass production (B) can be estimated as the dry matter production accumulated during the growing season or a calendar period under consideration. B can be related to the absorbed photosynthetically active radiation (R_{AP}) expressed in MJ m^{-2} and Light Use Efficiency ($\mathcal{E}$) of plants (g MJ^{-1}) using the relation developed by Monteith (1972).

$$B = R_{AP} \cdot \varepsilon \cdot 10 \quad \text{(Kg ha}^{-1}\text{)} \tag{6.1}$$

Plant leaves transmit and reflect solar radiation. Plant chlorophyll responds only to radiation in the 0.4 to 0.7 μm spectral regime, therefore only a fraction of the total broadband shortwave radiation (K) is available for photosynthesis (R_P). For 24 hours average conditions, the $R_P/K_{24}^{\downarrow}$ fraction is equal to 0.48 (Moran et al., 1995). Light interception occurs only in the case of green leaves filled with chlorophyll. R_{AP} can therefore be computed using Eq. (6.2).

$$R_{AP} = f \cdot \left(0.48 K_{24}^{\downarrow}\right) \quad \text{(MJ m}^{-2}\text{)} \tag{6.2}$$

The fraction f can be estimated directly from the normalized difference vegetation index, (NDVI) (Hatfield et al., 1984; Asrar et al., 1992):

$$f = -0.161 + 1.257\, NDVI \tag{6.3}$$

$\mathcal{E}$ for a particular crop is a constant if environmental conditions are non-limiting (Monteith, 1972). However, water availability and temperature can impact $\mathcal{E}$ significantly. SEBAL uses the equations first published by Field et al. (1995) to correct for the effect of heat variations (T_1, T_2) and soil moisture (W) on $\mathcal{E}$:

$$\varepsilon = \varepsilon' T_1 T_2 W \quad \text{(g MJ}^{-1}\text{)} \tag{6.4}$$

where $\mathcal{E}'$ is the maximum light use efficiency under optimal environmental conditions (g MJ^{-1}). T_1 and T_2 are heat variations based on the average temperature (T_{av} in $^{\circ}C$) and the average temperature during time step with maximum leaf area index (T_{opt} in $^{\circ}C$). W is a water scalar that is defined by the ratio of actual over potential evapotranspiration to describe the land wetness conditions. Bastiaanssen and Ali (2003) adopted the evaporative fraction (Λ) to define the water scalar of land mass. The evaporative fraction (Λ) is computed using Eq. 4.2 in Chapter 4.

For crops, the maximum light use efficiency $\mathcal{E}'$ varies with different plant types (C3, C4 and CAM plants) if there is no water shortage (Monteith, 1972; Steduto, 1996). The $\mathcal{E}'$ values have been provided for various crops including banana, maize, rice and sugarcane from various literature sources by Bastiaanssen and Ali (2003). Additional $\mathcal{E}'$ values from more recent literatures that also includes natural landscapes have been presented in Table 6.1 and 6.2. For natural landscapes, tropical forests have a maximum light use efficiency range of between 1.5 - 2.6 g MJ^{-1} (Heinsch et al., 2003; Ibrom et al., 2008). Shrublands and woodlands, and wetlands (high vegetation grass) have lower values and range between 0.8 - 1.6 g MJ^{-1} (Mobbs et al., 1997; Moncrief et al., 1997). During the nineties, radiation, water and carbon fluxes were measured on typical savannah sites in Niger as part of HAPEX-SAHEL (Goutorbe et al., 1997). The non-linear impact of soil moisture and $\mathcal{E}$ appeared to be an essential process. A good synopsis has been provided by Prince (1991) for applications with remote sensing.

6.2.3 Crop yield

The conversion from total biomass development to actual yield such as cereal grains varies with the harvest index (H_i) and the water content of the crop during the harvest (Donald and Hamblin, 1976).

$$Y_{act} = \frac{H_i}{1 - m_{oi}} \cdot \sum_{t=e}^{t=h} B_{day} = H_i^{eff} \cdot \sum_{t=e}^{t=h} B_{day} \quad (\text{Kg ha}^{-1}) \qquad (6.5)$$

where H_i^{eff} is the effective harvest index corrected using the optimal moisture content of the product after harvest.

Different crops have a specific harvest index range. The harvest index is also dependent on crop variety (Steduto et al, 2009; Yan and Wu, 2014). Management and agronomic practices such as the use of fertilizers can also increase crop yield as well as the harvest index to an optimal level of maximum productivity (Molden et al., 2010). H_i values for various crops have been used in the calibration process and are presented in Table 6.2.

6.2.4 Carbon sequestration

The accumulation of biomass is a result of Net Primary Production (NPP). The carbon assimilates are distributed across the various plant organs, including roots, shoots, tubers, stems, branches, leaves, flowers and grains. This partitioning is plant specific and the factors are empirically known. The ratio of above and below ground biomass production for natural landscapes has been measured for several biome types. The total biomass will reach a maximum value after the equilibrium between photosynthesis and natural decay due to seasons and age of the vegetation. Tropical rainforests with 40 m tall trees and ages of more than 150 years can reach up to 500 to 1,000 ton above ground biomass per hectare (Saatchi et al., 2011). Not unlikely, a significant amount of biomass is present in the soil. Ponce-Hernandez et al. (2004) estimated below ground biomass to be 0.25 – 0.30 of the above ground biomass using plant growth simulation models. Global maps of above ground biomass for the Pangani basin suggest values in the order of 15 to 40 ton ha^{-1}, depending on the type of ecosystem.

Litter and woody debris will be formed due to senescence and the age of vegetation. The decay of biomass will return carbon as carbon dioxide back into the atmosphere and partially into soil organic carbon. The typical carbon pools considered are (i) harvested wood, (ii) litter, (iii) dead wood and (iv) soil organic matter. This applies to all types of vegetation. The sequestration of carbon relates to various time scales: while litter and soil organic matter are generally conceived as short term carbon storage, harvested wood will have much longer time scales.

Carbon sequestration can be approximated as:

$$C = \xi \chi \sum B_{day} \tag{6.6}$$

where ξ [-] relates carbon pools to accumulate above ground biomass production and χ [-] relates total biomass production to above ground biomass production. The carbon pool does not take into account short term carbon storage such as plant leaves, litter that decay and may not change much over the years. The $\xi \chi$ factor is essentially the effective harvest index from biomass production for carbon sequestration.

6.2.5 Economic Water Productivity

The biophysical water productivity, $WP_{B/ET}$ (kg m^{-3}) is derived from plant biomass production and actual evapotranspiration (ET). The approach is widely adopted and has been used recently by Zwart et al. (2010) and Yan and Wu (2014).

$$WP_{B/ET} = \sum_{t=e}^{t=h} B_{day} \Bigg/ \sum_{t=e}^{t=h} ET \qquad (\text{Kg m}^{-3}) \tag{6.7}$$

The crop (yield) water productivity, $WP_{Y/ET}$ (kg m^{-3}) is derived from $WP_{B/ET}$ using the effective harvest index factor, H_i^{eff}, Eq. 6.8.

$$WP_{Y/ET} = WP_{B/ET} \times H_i^{eff} \tag{6.8}$$

The economic water productivity, $WP_{Ec/ET}$ (US$ m^{-3}) is derived from $WP_{Y/ET}$ and the net price of the product (P_n), Eq. 6.9.

$$WP_{Ec/ET} = WP_{Y/ET} \times P_n(i) \tag{6.9}$$

where $P_n(i)$ ($P_g(i)$ - $P_c(i)$) ($ kg^{-1}) is net farm gate price for crop i determined by subtracting the total production costs, $P_c(i)$ ($ kg^{-1}) from the gross farm gate price, P_g (i) ($ kg^{-1}) (Hellegers et al., 2009). The total production costs include both variable and fixed costs incurred during crop development. This approach of subtracting cost of production from the gross production relies on the fact that the value to a producer is exhausted by the summation of the values of the inputs required to produce it (Young, 2005).

The production cost was based on estimates provided by the agricultural extension office (Moshi, Pangani) for crops grown in the Lower Moshi irrigation scheme; rice, maize and vegetables. The farm gate prices for the banana were based on field data on the smallholder farms in the upper catchments of the river basin. The production costs included labour, fertilizers and pesticides costs, seedling, and harvesting and transportation expenses. Family labour and land is also included as variable costs using the prevailing labour rates and land rent costs. For commercial sugarcane, the economic water productivity was determined from the net benefits of sugar (sucrose) per unit of water consumption. The P_g is based on the wholesale price (2010) for white sugar of Tsh. 941 (US$ 0.67) per kg. The farm gate sugar price was slightly higher than the world average price (2010) for processed sugar (white) of US$ 0.60 per kg (LMC International, 2010). Data on the production costs for TPC sugarcane plantation was not available. The cost of production was therefore estimated from world averages. The production costs ranged between 45 - 70% and averaged 58% of the world sugar price for the 10-year period 2000 - 2009 (LMC International, 2010). The average value was used, disregarding the returns from bagasse (by-product) which constitutes 90% of the biomass.

Table 6.1. Net farm gate prices for crops under irrigation for 2008 - 2010 in Upper Pangani River Basin.

Crop	P_g Tsh kg^{-1}	P_c factor	P_n Tsh kg^{-1}	P_n US\$^{1} kg^{-1}
Rice (rough rice)	500	0.66	170	0.12
Maize	550	0.51	270	0.19
Vegetables (onions)	400	0.38	250	0.18
Bananas	200	0.15	170	0.12
Sugar (processed white)	941	0.58	395	0.28

^{1} *Exchange rate of 1US\$ to Tsh 1,409 (2010)*

For natural land cover the economic productivity may be inferred from the social and economic benefits of carbon sequestration. Here we assume that the value of ecosystem services is equal to the value of the net amount of additional carbon sequestered. This obviously a conservative assumption since there are other ecosystem services such as erosion control, nutrient recycling, waste treatment, provision of raw materials, habitat (genetic diversity) and cultural services that are provided by natural land cover (Costanza et al., 1997, 2014; De Groot et al., 2012).

According to the Interagency Working Group on Social Cost of Carbon (2009), the benefits of carbon sequestration varies from \$5 to \$65 per ton of carbon estimated from the social cost of carbon emission. Coffee farmers in Mexico received an amount of US\$ 13 per ton of carbon sequestered (AMBIO, 2010) while hydropower companies in Vietnam are being paid an amount of 20 Vietnamese Dong per Kilowatt hour (VNFF, 2014). These values are also consistent with the compensation market prices of between US\$ 5 to 30 per ton CO_2 equivalent - on emerging carbon offset markets (Batjes, 2012; Newell et al., 2012). A conservative fixed price estimate of US\$ 15 per ton of CO_2 is used in this study.

6.2.6 Additional datasets

For the spatial analysis of the economic water productivity, the boundary condition that outlines various land use types is required. Additional information on irrigation water uses is also required to assess the economic water productivity of blue water use.

Land use and land cover (LULC) types

This study made use of the LULC classification for the Upper Pangani River Basin in Chapter 3 (Kiptala et al. 2013a). The LULC types were derived using phenological variability of vegetation (from MODIS) for the same period of analysis, 2008 to 2010. 16 classes (including water bodies) were identified, of which smallholder maize (including supplementary irrigated) and shrublands are the two dominant classes, covering half of the area. The land use map is reproduced in Fig. 6.5a.

Irrigation water use

The actual evapotranspiration (ET) comprises of evapotranspiration from precipitation (P) and irrigation water use (Q_b). Q_b is supplied either through river water abstraction and/or groundwater. Q_b was estimated using the STREAM model (Chapter 5), a hydrological model developed specifically for the Upper Pangani Basin (Kiptala et al., 2014). The STREAM model utilized remotely sensed ET, soil moisture and P to simulate Q_b at 250-m and 8-day resolution. During low flows, Q_b consumed nearly 50% of the river flow in the river basin. Q_b estimates were comparable to field estimates of net irrigation with less than 20% difference (Kiptala et al., 2014).

6.2.7   Calibration and validation

The biomass parameters NDVI, evaporative fraction *(Λ)* and the incoming solar radiation ($K_{24}^{\downarrow}$) were derived from remote sensing. The other parameters T_1 and T_2 (Eq. 6.4), were computed using field based T_{av} and T_{opt} derived from the maximum LAI or NDVI in the plant growing season. Only two parameters were left for calibration: the maximum light use efficiency (ε') and the effective harvest index (H_i^{eff}).

The ε' values were adjusted within the experimentally verified parameter ranges from the literature. The actual yield data were used to derive the field based harvest index that was compared with the permissible H_i^{eff} range for moisture content of the product during harvest (m_{oi}). The actual yield data were taken from field measurements and yield records from main stakeholders in the Pangani River Basin.

Table 6.2 shows the experimentally verified ranges of ε' and H_i^{eff} for the four main crops grown in the Upper Pangani River Basin.

Table 6.2. Crop biomass parameters ranges for calibration.

Crop	ε' (g MJ⁻¹)	H_i^{eff} (kg kg⁻¹)	m_{oi} (kg kg⁻¹)	Sources
Rice	1.8-2.9	0.35-0.50	0.10-0.15	Casanova (1998), Boschetti et al. (2006), Boschetti et al. (2009)
Sugarcane	3.0-4.0	1.82-2.72	0.63-0.75	Varlet-Grancher et al. (1982), Bastiaanssen and Ali (2003), Waclawovsky et al. (2010)
Banana (bunch)	3.0-3.5	0.80-1.20	0.80-0.85	Nyombi (2010), Turner et al. (2008)
Maize	2.7-3.7	0.30-0.47	0.10-0.15	Varlet-Grancher et al. (1982), Maas (1988), Wiegand et al. (1991)

Since there are no actual yield data for the natural land cover, the average values of ε' as reported in the literature were adopted. The effective harvest index was also adopted from the literature on the basis of carbon sequestration. Carbon quantities were estimated at 0.43 - 0.55 of the above-ground biomass weight (Brown et al., 1989;

Kilawe et al., 2001; Namayanga, 2002; Ponce-Hernandez et al., 2004). An average carbon conversion factor of 0.5 to annual accumulated dry biomass was therefore adopted for this study (Table 6.3).

Table 6.3. Maximum light use efficiency and harvest index for natural land cover.

Vegetation type	$\mathcal{E}'$ (g MJ^{-1})	H_i^{eff} (kg kg^{-1})	Sources
Forest (tropical rain forest)	1.5 - 2.6	0.5	Heinsch et al. (2003), Ibrom et al. (2008)
Shrublands and woodlands	0.8 - 1.3	0.5	Mobbs et al. (1997), Molden et al. (2007)
Wetlands (high vegetation grass)	0.8 - 1.6	-	Li et al. (2012)

6.2.8   Uncertainty analysis of biomass production

Biomass production is related to vegetation growth and therefore has a temporal distribution due to seasonal variability. In remote sensing, the vegetation growth is accounted for by the phenological variability of NDVI over the cropping season or given time period (Kiptala et al., 2013a). Biomass production for a given land use type does not therefore follow a normal distribution. The nonparametric statistical inference is a technique to assess uncertainty for data that do not follow a normal distribution (Khan et al., 2006). The nonparametric bootstrapping technique was therefore used to estimate the confidence of mean biomass production. The pixel values of biomass were used as the sample population for the analysis. The bootstrapping draws random samples with replacement from the original population sample, each time calculating the mean or variance (Efron and Tibshirani, 1993). The process was repeated 1,000 times and a plot of the distribution of the sample means was made. The 95% confidence interval for the mean was determined by finding the 2.5[th] and 97.5[th] percentiles on the constructed distribution. The statistical software Minitab Inc (2003) was used in the analysis.

6.3   RESULTS AND DISCUSSIONS

6.3.1   Biomass production

Fig. 6.1a shows the spatial distribution of mean annual biomass production in the Upper Pangani River Basin based on calculations using Eq. (6.1) during a period of three years (2008-2010). The biomass growth covers 15 land use types (except water bodies) with the two main irrigation schemes, Lower Moshi (rice) and TPC (sugarcane) as shown in Fig. 6.1b and 6.1c respectively. The mean annual totals for the various LULC types are given in Fig 6.2.

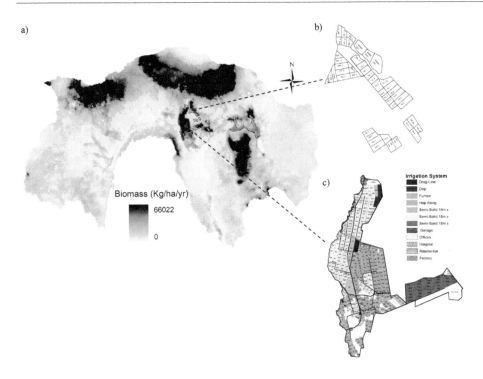

Fig. 6.1. (a) Spatial variability of biomass growth in Upper Pangani River Basin; (b) Lower Moshi irrigation scheme; (c) TPC sugarcane irrigation scheme.

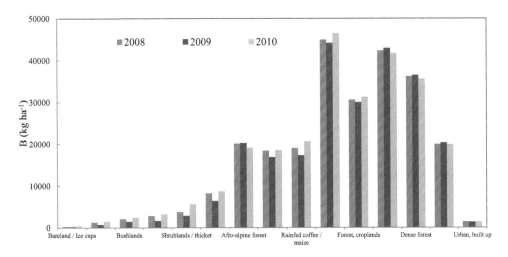

Fig. 6.2. Mean annual biomass production in the Upper Pangani River Basin for different land use types for the years 2008 – 2010.

The key drivers of the spatial and temporal variability are the precipitation, the biophysical characteristics of different LULC types and the inter-seasonal/intra-seasonal variation of the climate conditions during the period of analysis in the river basin. High biomass production is observed in the mountainous areas and in the main irrigation systems. The lower catchment areas that experience low rainfall have lower biomass growth. In the dry year of 2009, the biomass production was suppressed for most the LULC types in the lower catchment while enhanced in land cover types such as irrigated bananas and coffee, afro-alpine and dense forests and the wetlands that have sufficient water supply.

Crop biomass production

Table 6.4 shows the parameter values for light use efficiency and harvest indices from the calibration process for agricultural landscapes. The calibration process was based on actual field data for the four main crops in the Pangani Basin; rice, sugarcane, banana and maize.

Table 6.4. Calibrated crop biomass parameters from calibration against secondary data and their ranges reported in the literature.

Crop	$\mathcal{E}'$ (g MJ^{-1})	H_i (kg kg^{-1})	H_i^{eff} (kg kg^{-1})	m_{oi} (kg kg^{-1})
Rice	2.9	0.39	0.45	0.14
Sugarcane	3.5	0.69	2.20	0.68
Maize	2.7	0.30	0.35	0.14
Banana (bunch)	3.0	0.15	0.83	0.82

Rice

For rice, the calibration for biomass production was based on field data in the Lower Moshi Irrigation scheme (Fig. 6.1b). The Lower Moshi irrigation scheme has a command area of 2,300 ha for which 1,100 ha is used for paddy (rice) cultivation and the remainder mainly for maize and vegetable crops. Two rice varieties are cultivated, i.e. aromatic SAROS (TXD 306) and the non-aromatic IR64. The rice is grown as a monoculture with two crops per year to coincide with the *Masika* and *Vuli* rainfall seasons. However, due to water shortage, the rice in the middle to lower parts of the irrigation scheme is grown only during the *Masika* season, in rotation with maize or vegetables in the *Vuli* season. The growing cycle for the rice from germination to maturity ranges between 4-5 months. The irrigation scheme is organized in irrigation blocks of between 30-50 ha (5-8 pixels) which are further subdivided into small individual plots of approximately 0.3ha (Fig. 6.1b).

Rice yield sampling was done for rice grown in five upstream irrigation blocks where rice is grown in two seasons. The 1st season fell in the *Vuli* season (Dec 2008 to Apr 2009) and the 2nd season in the *Masika* season (May 2009 to Sept 2009). Rice yields were sampled at field level as rough rice (includes hull) and expressed as yield at 14% optimum moisture content at the rice milling plant (Table 6.4). The rice milling plant removes the outer hull and the bran layers from the rough rice. The whole grain white rice is estimated to be 67% by weigh to rough rice from field measurements at the Lower Moshi Rice Milling Plant. This is also consistent with the value of 65% mostly used in literature (Bouman et al., 2006).

Boschelli et al. (2006) showed that the maximum light use efficiency for high variety rice (flooded) has an averaged value of 2.9 g MJ^{-1} from agronomic field experiments using different rice cultivars. Traditional rice varieties had lower values of up to 1.8 g MJ^{-1}. Therefore, $\varepsilon' = 2.9$ g MJ^{-1} was adopted for the two high variety rice (TXD 306 and IR64) in the Pangani Basin. The average value will be slightly lower than the maximum when the limiting factors are taken in to consideration (see Eq. 6.4). The measured average ε for irrigated rice in Ebro delta, Spain for paddy rice was found to be 2.25 g MJ^{-1} (Casanova et al., 1998).

Table 6.5 shows the harvest indices from the field sampling data at Lower Moshi Irrigation scheme. The effective harvest index was calculated at 14% moisture content (moisture content requirement at rice milling plant).

Table 6.5. Actual yield to biomass sampling data for rice in the Lower Moshi irrigation scheme.

Irrigation block	Yield (ton ha^{-1} season^{-1}) 14% m_{oi}		Total yield rough rice (ton ha^{-1} yr^{-1})	Total B (ton ha^{-1}yr^{-1})	H_i^{eff} (rough rice)	H_i^{eff} (white rice)
	1st season 2009	2nd season 2009				
R 1-2	6.2	10.8	17.0	27.1	0.62	0.42
R 3-4	7.2	12.0	19.2	23.4	0.82	0.55
M 5-2	7.4	9.9	17.3	22.3	0.78	0.52
M 3-1	7.4	12.0	19.4	31.5	0.62	0.41
M. Kati	7.0	8.5	15.5	28.8	0.54	0.36
Average	7.0	10.6	17.6	25.9	0.68	0.45

The actual yield to biomass growth provides an effective harvest index (white rice) of 0.45 which is within the range of 0.35 - 0.50 provided in the literature (Table 6.2). However, the average harvest index of the two irrigation blocks (R 3-4 and M 5-2) where slightly above 0.50. This result may be attributable to higher yields generally associated with tropical rice cultivars (Bouman et al., 2007).

Sugarcane

In sugarcane production, the biomass calibration was based on actual yield data from the TPC sugarcane irrigation scheme (Fig. 6.1c). TPC sugarcane irrigation scheme covers an area of 8,480 ha of which 87% is under sugarcane plantation. The irrigation scheme is divided into 301 irrigations blocks of approx. 24 ha (4 pixels) using sprinkler (4,450 ha) or furrow (2,805 ha) irrigation. Drip irrigation was introduced in a few irrigation blocks (135 ha) as a trial. Sugarcane varieties N14, N19, N25 and N30 developed by the South Africa Sugar Research Institute (SASRI) are the dominant varieties cultivated. Other commercial varieties, including NCo 376, B52-313 and EA 70-97, are grown in small areas but have been replaced over time by SASRI varieties.

The cropping calendar for sugarcane ensures the harvesting of the cane during the relatively dry months from August to February. This also ensures that most of the crops are at development stage during the *Masika* high rainfall season. The sugarcane is harvested after 12 months at moisture content of 68%. The actual sugarcane yield ranges between 94 - 110 ton ha^{-1}, which is within the range for commercial sugarcane under irrigation (80 - 150 ton ha^{-1}) (Waclaworsky et al., 2010; Basnayabe et al., 2012).

The biomass production was analysed using an average ε' value 3.5 g MJ^{-1} derived from the ranges (Table 6.2). The biomass growth resulted in an average yield factor of 2.20, which was also within the range reported in literature (Table 6.4). The sugar content of 11% adopted from literature (Waclaworsky et al., 2010) provides an effective harvest index of 0.24 for sucrose. The harvest index is comparable with the value of 0.22 for sugarcane in Incomati basin, South Africa (Bezuidenhout et al., 2006; Hellegers et al., 2009).

Table 6.6. Actual biomass to actual yield for sugarcane in the TPC irrigation scheme.

Period	B (ton ha^{-1} yr^{-1})		Actual Yields (ton ha^{-1} yr^{-1})		H_i^{eff} (yield factor)
	Mean	STDEV	Mean	STDEV	
2008	44.9	10.9	95.7	23.9	2.13
2009	44.1	10.2	93.9	26.6	2.13
2010	46.4	10.3	110.3	27.7	2.38
Average	45.2	9.7	99.9	27.1	2.20

The results generally represent biomass production for commercial sugarcane as reported in the literature. The results are subject to uncertainty of 5% from the standard error on the actual yield data.

Maize

Maize is the dominant crop grown in the Upper Pangani River Basin. Smallholder maize is grown in the middle and upper catchments intercropped with beans, bananas and coffee in an expansive area of 294,200 ha. Irrigated maize is also grown in the Lower Moshi irrigation scheme in rotation with rice in an area of 59,800 ha. The average yield (2008-2010) for irrigated maize in Lower Moshi computed as 4.5 tons ha^{-1} was slightly lower than the world average (2010) of 5.1 tons ha^{-1} (FAO, 2013); but much lower compared to potential yield under irrigation of over 10 tons ha^{-1} (Hsiao, 2009; Jarmain et al., 2014). The average yields for rainfed maize were substantially lower and varied between 0 to 3 tons ha^{-1}. Despite the low yields of maize, the crop residue is also popular for silage and forage. The maximum light use efficiency of maize range between 2.7 - 3.7 g MJ^{-1}. The biomass production was analyzed using a lower margin ε' value of 2.7 g MJ^{-1} which resulted in an effective harvest index of 0.35 (Table 6.4).

Banana

Banana (bunch) is an important staple and cash crop grown mainly by smallholder farmers in the upper catchments of the Pangani Basin. The crop covers approximately 72,300 ha, intercropped with coffee or maize. The crop species grown is the East Africa highland banana (*Musa spp., AAA - EAHB*) under supplementary irrigation. The ε' for the East Africa highland banana range between 3.0 to 3.5 g MJ^{-1} (Nyombi, 2010; Turner et al., 2008). The average yield for banana is low at 15.5 tons ha^{-1} compared to the world average (2010) of 20 tons ha^{-1} (FAO, 2013). It is much lower than the potential which is reported to be over 60 tons ha^{-1} (Nyombi, 2010). A lower margin $\mathcal{E}'$ value of 3 g MJ^{-1} is used in the biomass production which provided a lower level effective harvest index of 0.83 from the ranges available from the literature (Table 6.2).

Natural ecosystems

For natural ecosystems, the maximum light use efficiency ($\mathcal{E}'$) is adopted from the literature (Section 6.2.2), as there was no field data on actual yield available for calibration. The average values of $\mathcal{E}'$ were used with the actual light use efficiency ($\mathcal{E}$) expected to be corrected by the spatial environmental factors that are highly variable due to the large topographical range in the river basin (Eq. 6.4). In the natural forest cover (dense forest and afro-alpine), an average $\mathcal{E}'$ value of 2.0 g MJ^{-1} was used from the permissible ranges for tropical rain forest. Similarly, average $\mathcal{E}'$ values of 1.0 and 1.2 g MJ^{-1} where adopted for shrublands and woodlands, and the wetlands respectively (Table 6.3). Implications of such selection in final results are discussed in the following section.

6.3.2   Uncertainty assessment for biomass production

The uncertainty for biomass production has been assessed through the confidence interval (CI) of the mean for each land use type. The lower and the upper bound confidence levels were estimated at 95% confidence limits using the bootstrap non-parametric technique (Efron and Tibshirani, 1993). The uncertainty has been greatly influenced by spatial coverage of the land use types within topographical or ecological zones. Urban and barelands had the highest uncertainty of 8% and 17%, respectively. The other land use types had lower uncertainties of less than 2% (Fig. 6.3 and Table 6.8). These results show that mean values of biomass production are representative for the given land use classes.

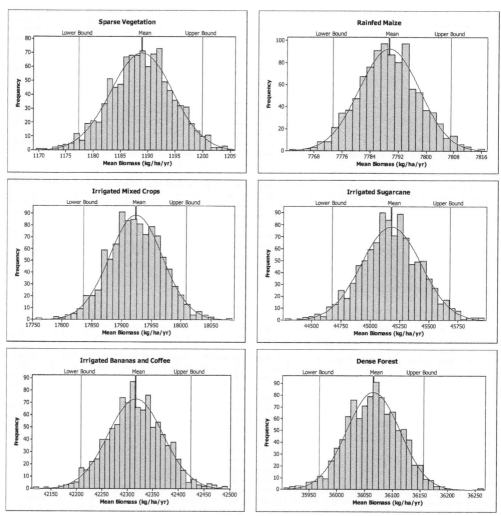

Fig. 6.3: Frequency distribution of the estimated annual mean biomass from bootstrap at 95% confidence limits for selected land use types in the Upper Pangani River Basin for period 2008–2010.

6.3.3 Water Yield

The mean annual water yields (P-ET) for 16 LULC are presented in Table 6.7. The water yields were derived P and ET computations in Chapter 4, Table 4.3.

The water supply is an important ecosystem service especially for forest ecosystem (Ford et al., 2011). The water yields and snow storage offers water regulation services to downstream catchment (Millennium Ecosystem Assessment, 2005; De Groot et al., 2012). From Table 6.7, the ice caps and afro-alpine forests generate significant water yields. The other natural ecosystems with large land mass: dense forest, bushlands and sparse vegetation also contribute substantial amounts of water flows to the river basin (Table 6.7). The water yields provide essential water flows for irrigation, wetlands and swamps, the water bodies (lakes, reservoirs) and hydropower for the Pangani river system.

Table 6.7. The water yields for various LULC types in the Upper Pangani River Basin for the period 2008 – 2010.

No.	Land use and land cover	Area km^2	Water yield (P-ET) mm yr^{-1}	10^6 m^3 yr^{-1}
1	Bareland/ice caps	100	*1,553*	*155*
2	Sparse vegetation	445	128	57
3	Bushlands	1,152	*162*	*187*
4	Grasslands/few croplands	1,517	61	93
5	Shrublands/thicket	3,509	*29*	*102*
6	Rainfed maize	2,942	-4	-12
7	Afro-alpine forest	257	*871*	*224*
8	Irrigated mixed crops	598	-17	-10
9	Rainfed coffee/irrig. banana	723	5	4
10	Irrigated sugarcane	89	-463	-41
11	Forest, irrig. croplands	556	-113	-63
12	Irrigated bananas, coffee	607	119	72
13	Dense forest	637	*186*	*118*
14	Wetlands and swamps	98	-647	-63
15	Urban, built up	8	202	2
16	Water bodies	100	-1,325	-133

Italics represent the LULC with the highest water contribution in the catchment

6.3.4 Water Productivity

Biomass and crop water productivity

The water productivity in terms of above-ground biomass and actual evapotranspiration ($WP_{B/ET}$) has been computed based on Eq. (6.7) (Table 6.8). The highest $WP_{B/ET}$ was realised in irrigated agriculture with sugarcane providing the highest $WP_{B/ET}$ of 4.4 kg m^{-3}. Dense and afro-alpine forests also attained high values of 2.4 and 1.4 kg m^{-3} respectively due to high biomass production. Land cover types with low biomass growth such as barelands or the sparse vegetation had the lowest $WP_{B/ET}$ values (Table 6.8).

The coefficient of variation (CV) of the mean (pixel) values of $WP_{B/ET}$ has been presented for each land use type. For natural land use types, a high level of CV (greater than 0.3) is related to the high level of heterogeneity of the land cover. For rainfed and irrigated croplands the CV is low to moderately high (0.1-0.3) compared to more homogeneous irrigation systems which have a CV of 0.05 (Zwart and Bastiaanssen, 2007). The biomass water productivity is converted into crop or yield productivity ($WP_{Y/ET}$) using their corresponding effective harvest indices.

Table 6.8. Average above ground biomass (B), actual evapotranspiration (ET) and water productivity ($WP_{B/ET}$) in Upper Pangani River Basin for the period 2008 – 2010.

Land Use and Land Cover No.		Annual B (kg/ha)			Annual ET (mm/yr)			WP (B/ET)	
		Mean	STDEV	CI[1]	Mean	STDEV	CI[1]	Mean (kg/m³)	CV[2]
1	Bareland/ice caps	319	538	27	643	653	32	0.05	1.3
2	Sparse vegetation	1,189	477	11	586	172	4	0.20	0.6
3	Bushlands	1,999	1,017	15	669	312	5	0.30	0.4
4	Grasslands/few croplands	2,550	652	8	630	223	3	0.40	0.4
5	Shrublands/thicket	4,100	1,209	10	756	85	1	0.54	0.3
6	Rainfed maize	7,789	1,870	17	789	221	2	0.99	0.3
7	Afro-alpine forest	19,803	5,529	171	1,429	309	9	1.39	0.2
8	Irrigated mixed crops	17,923	4,133	86	905	207	4	1.98	0.3
9	Rainfed coffee/maize	18,973	4,352	80	1,022	261	5	1.86	0.2
10	Irrigated sugarcane	45,175	9,651	501	1,035	212	11	4.36	0.2
11	Forest, croplands	30,612	5,250	109	1,228	250	5	2.49	0.2
12	Irrigated bananas, coffee	42,316	5,239	108	1,330	156	3	3.18	0.1
13	Dense forest	36,065	4,819	94	1,517	144	3	2.38	0.1
14	Wetlands and swamps	20,039	4,415	219	1,291	267	13	1.55	0.2
15	Urban, built up	1,409	327	57	774	80	14	0.18	0.6

[1]*CI - confidence interval of mean at 95% confidence limits*
[2]*CV - coefficient of variation of the mean (pixel) values of water productivity*

For sugarcane, the estimated $WP_{B/ET}$ of 4.36 kg m⁻³ was within the range reported in literature (3.5-5.5 kg m⁻³) (Thompson, 1976; Olivier and Singels, 2003; Hellegers et al., 2009; Carr and Knox, 2011). The average $WP_{Y/ET}$ of 9.6 kg m⁻³ sugarcane (1.1 kg m⁻³ sucrose) compares well with the crop water productivity data reported for irrigated sugarcane of between 4.0-11.1 kg m⁻³ in India and 7.4 kg m⁻³ in Thailand (Mainuddin and Kirby, 2009). The average yield for sucrose (1.1 kg m⁻³) is also consistent with the estimates of 1.1-1.3 kg m⁻³ in the Incomati Basin (Hellegers et al., 2009). There was no significant difference between the water productivities of furrow and sprinkler irrigation systems. Drip irrigation systems produced lower yields (average of 92 tons ha⁻¹) with a lower actual ET usage, thus resulting in a statistically similar $WP_{Y/ET}$ (9.6 kg m⁻³) compared with the other irrigation technologies.

$WP_{B/ET}$ for rice of 1.5 kg m⁻³ is within the range (0.6-1.6 kg m⁻³) reported in the literature (Zwart and Bastiaanssen, 2004; Bouman et al. 2006; 2007; Mainuddin and Kirby, 2009; Zwart and Leclert, 2010). For irrigated bananas, the average $WP_{B/ET}$ and $WP_{Y/ET}$ is 3.2 and 2.8 kg m⁻³ respectively. The average $WP_{B/ET}$ for irrigated banana is in the range of 1.4 – 5.5 kg m⁻³ in Nico Coelho irrigation scheme, Pernambuco, Brazil (Bastiaanssen et al., 2001). The $WP_{Y/ET}$ is also consistent with estimates of 2.8 kg m⁻³ for the banana crop in the Incomati Basin (Hellegers et al., 2009). The water productivity $WP_{B/ET}$ for rainfed maize (with supplementary irrigation) is 1.0 kg m⁻³ while for irrigated maize it is 2.0 kg m⁻³. The water productivity translates to $WP_{Y/ET}$ of 0.35 and 0.70 kg m⁻³ for rainfed and irrigated maize crops. The low productivity for rainfed maize is explained by the low yields that are generally associated with water stress due to long dry spells during the growing seasons. The result is consistent with a recent study that showed that the water productivity for irrigated agriculture is twice that of rainfed agriculture in the Indus Basin (Karimi et al., 2013b). In general, the maize productivity $WP_{B/ET}$ is within the range reported in literature of 1.1-2.7 kg m^{-3} (Zwart and Bastiaanssen, 2004). The yield productivity $WP_{Y/ET}$ was also consistent with the range of 0.40 to 0.70 kg m^{-3} for irrigated maize in Mkoji, Great Ruaha river basin in Tanzania (Igbadun et al., 2006).

For natural land cover, the $WP_{B/ET}$ is high for natural tropical forest because of the high biomass production associated with favourable rainfall throughout the year. $WP_{B/ET}$ for dense forests averaged 2.4 kg m^{-3} which was within the range of between 2 - 3 kg m^{-3} for closed forest in the Nile Basin for year 2007 (Molden et al., 2009). The relatively high $WP_{B/ET}$ value for the wetlands and swamps of 1.6 kg m^{-3} is consistent with the values found for natural wetlands in the Nile Basin of 1.5 kg m^{-3} (Molden et al., 2009) and can be explained by year round access to water. The $WP_{B/ET}$ for shrublands, bushlands, sparse vegetation and barelands, located in the lower parts of the Upper Pangani, are lower than 0.5 kg m^{-3}, mainly because of low rainfall. In natural grasslands, the average biomass production of 2.6 tons ha^{-1} represents a $WP_{B/ET}$ of 0.4 kg m^{-3} (Table 6.8). The water productivity is low compared to good pastures or forage (Alfalfa) grown for commercial purposes under irrigation which range between 1.0-2.6 kg m^{-3} (Grimes et al., 1992). The low productivity may be attributed to low rainfall and possibly combined with overgrazing.

The relationship between average B and ET for various landscapes is presented in Fig. 6.4.

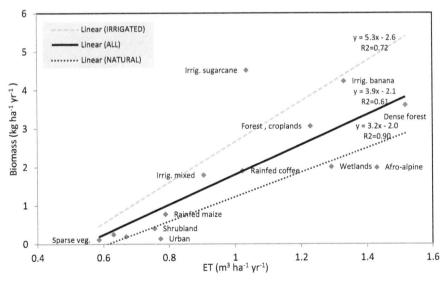

Fig. 6.4. Variation between biomass and actual evapotranspiration (ET) for agricultural and natural landscapes using average values for period 2008 - 2010 in the Upper Pangani River Basin.

There is a moderately good linear relationship between the B and ET in irrigated ($R^2=0.72$) and even stronger in natural ($R^2=0.90$) landscapes. It is observed that biomass production in irrigated agriculture is more enhanced than in rainfed agriculture and natural landscapes. Most of the irrigated crops (maize, banana, and vegetables) are under supplementary irrigation where little blue water is added in critical times to supplement rainfall. This improves the water availability for the crops. In commercial sugarcane (fully irrigated), irrigation combined with soil fertility management provides optimal conditions for plant development. Fig. 6.4 shows the scope for increasing water productivity. Additional water storage (through water harvesting) and the subsequent water use by rainfed systems will result in a significant increase in biomass production. This can be observed by comparing rainfed maize and irrigated mixed crops (maize, beans and vegetables) where the biomass production increases by 130% with only 15% difference in ET. It is noteworthy also that different kinds of plants (C3, C4 and CAM plants) have different water use efficiency in terms of B and ET in its positioning in Fig.6.4. As such C4 crops such as sugarcane and maize are more water efficient than C3 crops such as rice (Molden et al., 2007). Fig. 6.4 can also provide implicitly (from water use) the shadow value or the opportunity cost of keeping the natural ecosystems in good conditions. These values are baseline data that can be used for detailed environmental valuation and modelling.

Fig. 6.5 presents that land use and the corresponding water productivities in the Upper Pangani Basin.

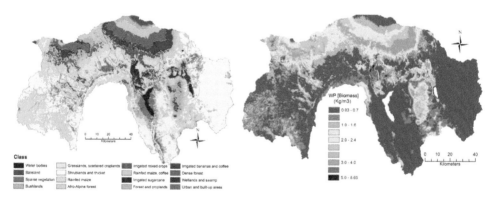

Fig. 6.5 (a) Land use and land cover map (Kiptala et al., 2013a) (b) Water Productivity (Biomass) for period 2008 – 2010

Economic water productivity

The economic water productivity was computed from the crop yields and the net production values using Eq. 6.9. Table 6.9 provides the yield and the corresponding economic water productivity for the main crops in the basin.

Table 6.9. Crop yield and economic water productivity for main agricultural crops in Upper Pangani River Basin.

Land use land cover type	Crops[1]	$WP_{Y/ET}$ (kg m^{-3})	$WP_{Ec/ET}$ ($ m^{-3})
Irrigated mixed crop	Rice	1.5	0.18
Irrigated mixed crop	Irrigated maize	0.7	0.13
Rainfed coffee/maize	Rainfed maize	0.35	0.07
Irrigated mixed crop	Vegetables	1.6	0.29
Irrigated bananas, coffee	Bananas	2.6	0.31
Irrigated sugarcane	Sugarcane (commercial)	9.6 (1.1[2] Sucrose)	0.31

[1]The crops were selected from the pixel locations
[2]Biomass to sugar (sucrose) harvest index of 0.24

$WP_{Ec/ET}$ values for irrigated crops show that sugarcane and bananas have a higher economic water productivity (0.31 $ m^{-3}) compared to those of vegetables (0.29 $ m^{-3}), rice (0.18 $ m^{-3}) and maize (0.13 $ m^{-3}). The average $WP_{Ec/ET}$ for rice was within the range of 0.10-0.25 $ m^{-3} reported in the literature (Mainuddin and Kirby, 2009; Molden et al., 2010). Rice and maize have relatively stable local markets and are the preferred crops in the Lower Moshi irrigation scheme. The banana and vegetable markets are controlled by middlemen and the market is volatile. The P_g for banana and vegetable crops was about 44% of the retail prices of the produce at the local markets. The high difference is mainly attributed to the high transport and brokerage costs.

The economic water productivity for commercial sugarcane of 0.31 $ m^{-3} was higher in 2008/10, compared to Incomati basin (2004/05) of 0.20 $ m^{-3} (Hellegers et al., 2009). Since the biomass and crop yield productivities were comparable, the difference can be attributed to the high sugarcane prices between the two periods (LMC International, 2010).

For pastures (grasslands and scattered croplands), the economic water productivity was assessed using the biomass production that is grazed or harvested mainly for livestock. The assumption here is that the grass consumed by the livestock could otherwise have been purchased as dry feeding forage (hay) in assessing the economic water productivity. Using an effective harvest index of 0.70 for dry grass at 15% moisture (harvest index of 1 and field moisture of 50%); the $WP_{Y/ET}$ becomes 0.28 kg m^{-3}. The market price for hay (dry grass) is approx. 0.18$ kg^{-1} (20 kg hay market price averaged Tsh. 5,000). This implies thus that the $WP_{Ec/ET}$ could be approximately 0.025 $ m^{-3} using a cost of production factor of 0.5. The productivity (natural grasslands) is approximately three times less than rainfed maize (0.07 $ m^{-3}) despite the small difference (20% in average) in water usage (ET). However, the grassland's water yield is higher than rainfed maize (Table 6.7), thus providing additional ecosystem services to the catchment.

The dense forest and afro-alpine forest with an annual average biomass growth of 20 tons ha^{-1} yr^{-1} and 36 tons ha^{-1} yr^{-1} (Table 6.8) has carbon storage of between 10 - 18 tons C ha^{-1} yr^{-1}. The economic potential for the carbon storage in forest would therefore range between 150-270 US$ ha^{-1} yr^{-1}. Using the water productivity of carbon yield (from biomass), $WP_{Y/ET}$ (CO_2) of 0.7-1.2 kg m^{-3}, the $WP_{Ec/ET}$ translates to a range of between 0.01 - 0.02 US$ m^{-3}. The $WP_{Ec/ET}$ is low (about 15 times less) compared to irrigated agriculture despite high $WP_{B/ET}$ values (Table 6.8). This is mainly attributable to the low carbon market prices. The low cost of carbon market prices was also noted by Batjes (2014) as the greatest hindrance for quicker implementation of the CO_2 mitigation measures in agriculture. However, there are prospects that the situation is poised to change with emerging markets for carbon trading where countries (through their industries) trade to meet their CO_2 obligations specified by the Kyoto Protocol.

Shrublands and bushlands also provides for carbon storage. The annual biomass production ranges between 2 - 4 ton ha^{-1} and the carbon storage will thus range between 0.5 - 1.0 ton C ha^{-1} yr^{-1}. However, other provisioning and regulatory services for these biomes such as providing grazing fields for wildlife and food for local populations, as well as water regulation, add significant value to this land use (Costanza et al., 1997).

Table 6.10 presents the yield and economic water productivity for the natural landscapes based on hay and carbon sequestration. The economic value for carbon is expected to increase by 25 – 30% to account for below ground biomass production.

Table 6.10. Yield and economic water productivity for natural landscapes in Pangani River Basin.

Land use land cover	Crops[1]	$WP_{Y/ET}$ (Kg m^{-3})	$WP_{Ec/ET}$ ($ m^{-3})
Grassland	Hay	0.28	0.025
Dense/afro-alpine forest	Carbon storage	0.7-1.2	0.01-0.02
Shrublands and bushlands	Carbon storage	0.15 – 0.27	< 0.004

Fig. 6.6 presents the spatial crop yield and the economic water productivities in the Upper Pangani Basin. Both productivities accounted for CO_2 storage for forest and shrulands, hay for grasslands and the harvestable yield for irrigated and rainfed agriculture.

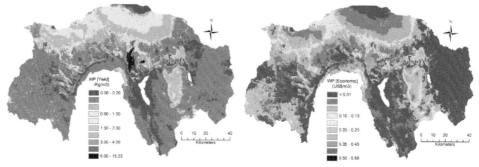

Fig. 6.6 (a) Water Productivity (Yield) (b) Water Productivity (Economic) for period 2008 - 2010.

Apart from carbon storage, natural biomes provide other valuable services that may relate to ecological production. Water yields in natural landscapes are consumed in agricultural landscapes. Some of them are not monetized. Agricultural landscapes and urban areas also have some disservices as well (Van Berkel and Verburg, 2014; Gómez-Baggethun and Barton, 2013). Fanaian et al. (2015) provides a good and systematic overview of how to undertake local economic valuation of ecosystem services by way of assessing tradeoffs of alternative resource use in a river system. In this chapter, literature is referred to infer the extent of the value of the natural ecosystem services provided by the natural ecosystems part from the CO_2 sequestration considered in Table 6.10. De Groot et al. (2012) provided the ranges of economic values (2007 prices) based on over 3,000 peer reviewed studies, some in the tropical climate. Of interest are the tropical forest, woodlands, grasslands, inland wetlands and fresh water lakes that constitute much of the natural ecosystems in the Upper Pangani River Basin.

The economic value for provisioning services for tropical forests is estimated at 2,695 US\$ ha^{-1}yr^{-1} from 96 case studies at a standard error of 13%. From the water use for dense and afro-alpine forest in the Upper Pangani River Basin, the water productivity equals 0.19 US\$ m^{-3}, which is 10 times greater than the derived value for carbon storage (Table 6.10). Furthermore, the water value could significantly increase to 0.34 US\$ m^{-3} (5,264 US\$ ha^{-1} yr^{-1}) if the other ecosystem services (water and climate regulation, erosion control, nutrient recycling) were also considered.

The shrublands provide primarily food and pasture to livestock, wildlife and even local communities in form of plant leaves and fruits. From section 6.2.3, this part of biomass was not accounted for in carbon sequestration. From literature, the economic value of shrublands is estimated at 1,305 US\$ ha^{-1} yr^{-1} based on 21 case studies with standard error of 4% for which 90% of the value is derived from food (De Groot et al., 2012). The assessment is consistent with the shrublands areas in the Africa savanna including Pangani, which hosts wide variety of wildlife, and pastoralists keeping cattle. Considering the water use for the shrublands in the Upper Pangani River Basin, the water productivity equals 0.17 US\$ m^{-3}. Habitat (genetic diversity) and cultural services of the shrublands increases further the water productivity to 0.38 US\$ m^{-3}. On the other hand, bushlands provide ornamental and generic resources such as medicine to the local communities (Costanza et al., 1997) which is estimated at 253 US\$ ha^{-1} yr^{-1} (De Groot et al., 2012), equivalent to a water value of 0.04 US\$ m^{-3} in the Upper Pangani River Basin. Considering habitat services, the economic value of the bushlands would increase to 1,588 US\$ ha^{-1} yr^{-1} or a water value of 0.24 US\$ m^{-3}. These water values for both the shrublands and bushlands are significantly much higher than the economic benefits from carbon sequestration (<0.004 US\$ m^{-3}) given in Table 6.10.

Inland wetland is known to provide the greatest value in ecosystem services in the form of water and disturbance regulation, waste treatment, food production and recreation. Although most of the functions are not directly related to biomass, its total economic value (2007 prices levels) was estimated at 25,682 US\$ ha^{-1} yr^{-1} (De Groot et al., 2012). The valuation is consistent with earlier estimates (1994 price levels) of 15,000 US\$ ha^{-1} yr^{-1} (Constanza et al., 1997). The provisioning services that relate to ecological production (biomass) such as food, pasture and raw materials is estimated much lower at 1,659 US\$ ha^{-1} yr^{-1}. The average value is based on estimates from 168 case studies at 11% standard error. The economic value from the provisioning services is 0.13 US$ m^{-3} from water use (evapotranspiration) of 1300 mm yr^{-1}. The water value would increase to 2.0 US$ m^{-3} if all the other regulating and habitat services (2007 price levels) were considered.

The fresh water lakes also provide substantially high value in terms of water supply, water treatment, food (fish and grazing land) and recreation services. The water value from 15 case studies at 17% standard error is estimated at 1,914 US$ ha^{-1} yr^{-1} (De Groot et al., 2012). In the Pangani River Basin, the Lake Jipe and Lake Chala provide these provisioning services. Considering the evaporation rate of 2000 mm yr^{-1}, the water value equals 0.10 US$ m^{-3}. Costanza et al. (1997) argues that additional water regulation services provided by the wetlands should also be added to the freshwater lakes. Thus, recreational services would increase the water value of the fresh water lakes to 0.4 US$ m^{-3}. However, these natural biomes are located close to each other in the Upper Pangani River Basin and the water regulation may only be offered by one or partly by all.

The comparison of the results for natural landscapes derived in Table 6.10 (based on CO_2 and pastures) against literature reviewed showed that the biophysical analysis based on carbon storage and pastures only, significantly underestimates the total economic value of the natural ecosystem services. However, a local ecosystem valuation needs to be undertaken to explicitly derive the total value of ecosystem services generated by each natural biome. This will reduce the chances of avoiding double counting especially for parallel ecosystem services supported by nearby natural landscapes. As such the biophysical data and the baseline water values provided in this study provide boundary conditions for further detailed economic valuation.

Economic water productivity for irrigation water use

Crop water use (actual ET) comprises of both precipitation (P) and net irrigation (blue water) withdrawal (Q_b). Table 6.11 shows the percentage of Q_b to the total actual water use (ET) for the main irrigated crops in the Upper Pangani River Basin extracted from Kiptala et al. (2014). The economic water productivity in terms of blue water ($WP_{Ec/ETb}$) is derived by dividing the total economic water productivity ($WP_{Ec/ETb}$) by the blue water use factor (Q_b/ET).

Table 6.11. Economic water productivity for main irrigated crops expressed in terms of net irrigation water use for the period 2008 - 2010.

Crop	Q_b (%)	$WP_{Ec/ET}$ (US$ m^{-3})	$WP_{Ec/ETb}$ (US$ m^{-3})
Sugarcane	44	0.31	0.70
Rice	36	0.18	0.50
Vegetables	24	0.29	1.21
Bananas	21	0.31	1.48
Maize	17	0.13	0.76

Economic (blue) water productivity ($WP_{Ec/ETb}$) showed higher values for smallholder crops (banana, vegetables and maize) using less Q_b compared to fully irrigated rice and sugarcane. During periods of physical water scarcity, the opportunity cost for blue water (scarce) resources is much higher than that of precipitation.

6.4 DISCUSSION AND CONCLUSION

This chapter presents the spatial water productivity in terms of bio-physical (biomass and yield) and economic benefits for different LULC in the Upper Pangani River Basin in Eastern Africa. The methodology applied remotely sensed data to estimate accumulated biomass growth and field based data to estimate yield and economic indices. The study also relied on literature for crop yields and ecosystem services to estimate spatially explicit water productivity and water value in a data scarce and heterogeneous landscape in Africa. The results were based on three years of analysis (2008 (wet), 2009 (dry), 2010 (average)). Such water productivity indices generated can inform strategies for the optimal allocation of water in river basins such as the Upper Pangani.

In agriculture, irrigated lands achieved the highest water productivity in terms of biomass, crop yield and economic productivity. The productivity for sugarcane and rice is high and within the ranges reported in the literature. In supplementary irrigated and rainfed systems, the productivity was moderate but significantly lower than the potential reported in literature. In natural ecosystems, natural forest and wetlands that have access to abundant water resources achieved a relatively high biomass production compared to the expansive shrublands and sparse vegetation. However, the economic productivity for the natural forest and wetlands was low when computed using biomass derived CO_2 storage only.

We found a linear relationship between biomass (and yield) and ET in both agricultural and natural landscapes. As expected, biomass production in irrigated (and supplementary irrigated) agriculture was higher than in rainfed agriculture and natural ecosystems in Upper Pangani. This can be explained by better plant water and soil fertility management in the irrigated landscapes, and relatively low rainfall in the rainfed landscapes. The relationship provides scope for improved water productivity especially for rainfed systems with little irrigation and good agronomic practices. Molden et al. (2007; 2010) provides various ways of increasing productivity for rainfed agriculture and crops under supplementary irrigation (see also Van der Zaag, 2010; ILRI, 2014).

The pixel results for biomass water productivity of a given land use class showed large variations. The coefficient of variation (CV) is high for natural and conserved land uses (0.3-1.3). This is mainly due to high topographical ranges that influence the environmental factors (rainfall, temperature). For agricultural land uses, the CV is lower (0.1-0.3), but nevertheless still higher than reported for large scale irrigation systems. The result can be explained by the high levels of intercropping that exist within the land use types. Even so, the spatial analysis provided further scope for increased water productivity in agricultural plots that have low biomass growth.

The economic productivity in terms of irrigation water use (Q_b) was higher for the smallholder banana and maize under supplementary irrigation compared to the fully irrigated sugarcane and rice. In situation of physical water scarcity, the opportunity cost for blue water (scarce) resources is much higher than that of precipitation. It would thus be more prudent to allocate river water to the supplementary irrigated crops than to fully irrigated crops that are grown during the dry season.

The low market price for carbon sequestration was observed to lower the water productivity for the natural forest ecosystems. The market price was observed to be even lower than the social costs of carbon emission in the global market (Interagency Working Group on Social Cost of Carbon, 2009; Batjes, 2012). Nevertheless, the price is expected to increase significantly in the near future with expected carbon trading following the Kyoto Protocol (Carbon Market Watch, 2014). Additional provisioning ecosystem services for natural land cover inferred from the literature increased substantially the economic water productivity especially for the wetlands, natural forest and the shrublands. Advanced techniques for localized environmental and social ecosystem valuation for natural environment exist (FAO, 2004), but these techniques require specific expertise and are quite costly (Hermans and Hellegers, 2005). However, the biophysical data and baseline water values derived here can therefore provide boundary conditions for further detailed economic or environmental modelling.

The social water value to the local populations is also a key factor for consideration by policy makers when interpreting water productivity indices. Banana, maize and rice provide staple food for local livelihoods and income thus enhancing food security and social well-being. On the other hand, sugarcane grown by the TPC Company has a relatively high economic productivity, may not provide high benefits to local livelihoods. Trade-offs therefore has to be made between attaining high economic water productivities and social equity.

It is clear from the spatial water productivity maps that biomass production is closely correlated with economic water productivity. Water yields, carbon credits and other ecosystem services provide insight into the water value society attaches to a certain cultural and natural land use activity. Standard PES also relates with efforts to enhance water yields, prevention of soil erosion and carbon sequestration amongst others. A holistic approach that involves the improvement of both the biophysical and economic factors through sustainable basin-wide interventions can therefore result into higher economic water productivity at lower social and environmental cost. Such a comprehensive water productivity analysis will allow policy makers to quantify the foregone economic benefits for allocating water for socio - environmental gains. Moreover, the physical water productivity can inform long term basin strategies based on future market trends and socio-economic scenarios. The remotely sensed estimate of the economic water productivity that includes natural ecosystems would therefore provide vital information for sustained green growth and socio-economic development in many African river basins.

Chapter 7

MULTI-OBJECTIVE ANALYSIS OF GREEN-BLUE WATER[5]

The concept of integrated water resource management (IWRM) attempts to integrate all relevant elements related to water resources. Different tools exist that can inform sound IWRM plans and identify trade-offs. One such tool is multi-objective analysis using integrated hydro-economic models (IHEM). However, IHEM mainly deals with the optimization of river flow (blue water) in a river basin and does not incorporate water used in the landscape. This chapter connects these two elements by linking a distributed model of green and blue water uses in the upper catchment, with the mainly blue water uses in the lower part of a basin. As such, it allows for basin wide analysis of water use. The analysis focuses on maximizing three primary objectives: i) hydropower production, ii) supplementary irrigation where crop water requirement is met by both precipitation (green water) and river abstraction (blue water), and iii) fully irrigated agriculture where all crop water requirements are met only by the blue water. The analysis also considers five socio-environmental objectives, and is conducted for a wet, a dry and an average year.

The results show that agricultural water use (supplementary and fully irrigated) achieves relatively high water productivity and competes with all the other objective functions. The guaranteed hydropower at 90% reliability (firm energy) favours constant flow conditions throughout the year, which then competes with the environment that requires both high and low flows. Water abstraction for smallholder irrigation and urban use deprive downstream hydropower and the environment of water. The study shows that improving rainfed cropping (maize) through supplementary irrigation during the rainy seasons has a slightly higher marginal water value than full scale irrigation (sugarcane). The developed methodology may be applicable in other river basins with predominantly green water use upstream and blue water use downstream.

---

[5] This chapter is based on the paper: Kiptala, J.K., Mul, M.L., Mohamed, Y.M., Van der Zaag, P., 2016b. Multi-objective trade-off analysis of green-blue water uses in a highly utilized river basin in Africa. Submitted to *Journal of Water Resources Planning and Management, American Society of Civil Engineers (ASCE)*.

7.1 INTRODUCTION

Water is an important natural resource for all forms of life and it forms the backbone for economic productivity and social wellbeing in many parts of the world. With growing demands for water, it is becoming increasingly challenging to satisfy those needs. Competition between different water uses and between upstream and downstream use is therefore increasing. Many river basins are overexploited, and the capacity to meet the different social demands is decreasing (Mostert et al., 1999; Molden et al., 2007).

In Africa, over 60% of the total population relies on water resources that are limited and highly variable (UNEP, 2010). About 75% of the continent's cropland is located in arid and semi-arid areas, where irrigation can greatly improve productivity and reduce poverty (Smith, 2004; Vörösmarty et al., 2005). There has been an increased focus on the development of multipurpose reservoirs during recent years. These dams have enabled better water management that have significantly increased economic benefits to river basins. However, many large storage projects worldwide are failing to produce the level of benefits that provided the economic justification for their development (WCD, 2000; Ansar et al., 2014). This is due to the overstated benefits portrayed in the feasibility documents among other factors. In addition, there are social and environmental externalities that include the displacement of communities and the alteration of the hydrology that cause disbenefits to downstream communities and the natural environment. The vulnerability of the often poor riparian population who rely on ecosystem services generated by the natural environment is increased due to these developments (Malley et al., 2007).

According to Postel (1992), the main thrust of the management of river basins is finding ways of turning these potential conflicts into constructive cooperation, and to turn what is often perceived as a zero–sum predicament, in which one party's gain is another's loss, into a win–win proposition. While the physical water is a finite resource, the quantity of water resources available can be influenced by management decisions. This requires a broad perspective in the management of water resources; looking at maximizing benefits (Sadoff and Grey, 2002) and allocating sufficient water to the environment to secure ecosystem services (Rood et al., 2005; Konrad et al., 2012).

Integrated hydro-economic modelling (IHEM) tools have been developed to integrate economic efficiency and equity objectives in river basins (Ward et al., 2006; Ringler and Cai, 2006; Pulido-Velazquez et al., 2008). These studies have used a range of deterministic and stochastic single to several objective problem formulations. In recent studies, advanced multi-objective optimization algorithms that rely on Pareto-optimal curves or surfaces have been developed (Kaspryk et al., 2009; Kollat et al., 2011; Reed et al., 2013). These methodologies require high computational effort, and use super computers and parallel computing techniques (Hurford and Harou, 2014). However, all these studies focus solely on the blue water use in a river system.

Solving the challenge of water resources management is not only about blue water allocation (water in rivers, aquifers, lakes, reservoirs) but includes green water use (soil moisture from precipitation). Green water use through rainfed agriculture generates most of the food in Africa, yet the productivity per ha remains low. A key strategy to upgrade rainfed agriculture is investment in supplementary irrigation to bridge dry spells (Falkenmark and Rockström, 2006). Improved water productivity in rainfed systems can be significantly increased with little extra water use (Molden et al., 2007). However, an increase in green water use will inevitably result in a decrease in blue water availability. Water resources management decisions should therefore reconsider the predominant focus on blue water to the full water balance that also includes green water use.

In this study, a multi-objective exploration of trade-offs in competing water uses is analysed for the Pangani River Basin in East Africa. Enhanced green water use for improved rainfed and supplementary irrigated agricultural systems is integrated into the IHEM. The following section outlines the Pangani river system, followed by the IHEM model set-up for Lower Pangani hydro-system and the description of model scenarios. The results and discussion are presented in Section 7.4, and finally the conclusions are given in Section 7.5.

7.2 PANGANI RIVER SYSTEM

The Pangani River system starts at the mountains of Mt. Meru, Mt. Kilimanjaro and the highlands of the Pare and Usambara mountains and runs through the semi-arid middle course into the Pangani estuary and empties in the Indian Ocean (Fig. 7.1). The Upper Pangani River Basin, defined as the catchment area upstream of Nyumba ya Mungu (NyM) reservoir, is the main source of water (*blue water*) for the Lower Basin (Kiptala et al., 2013a; 2013b). Rainfall (300-800 mm yr^{-1}) has a bimodal pattern where long rains are experienced in the months of March to May (Masika season) and the short rains in the months of November to December (Vuli season).

The water resources from the main upstream sources of Kikuletwa and Ruvu rivers provide water for hydropower and irrigation and also provide essential environmental flows to maintain key ecosystem services such as the Kirua swamp in the lower catchments. The water resources are supplemented by the Mkomazi, Soni and Luengera rivers along the river system that stretches for over 500km. The three hydroelectric power stations (HEP) in the lower basin; NyM (8MW), Hale (21MW) and New Pangani Falls (NPF) (68 MW) together contribute 17% of total installed hydropower generation capacity to the Tanzania national grid (Table 7.1 presents the key features of these HEPs).

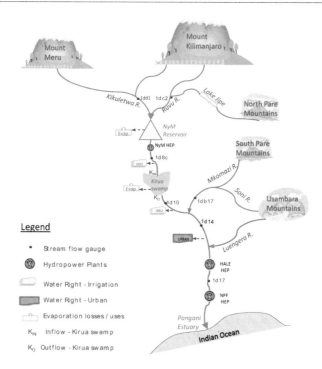

Fig. 7.1: Schematic layout of the Pangani River System

NyM reservoir with a storage capacity of 1.14×10^9 m^3 regulates the river flow downstream. The dam has dramatically changed the downstream river regime; from the bi-annual flooding of the floodplain to a fully controlled flow in the river channels. This has led to the reduction of Kirua swamp from an area of 852 km^2 to 10 km^2 (PBWO/IUCN, 2008). The Lower Pangani River basin also represents a major potential area for irrigated agriculture. According to records at Pangani Basin office, by 2010, a total flow of 3.12 m^3 s^{-1} has been issued as water rights for smallholder irrigation systems in the Lower Pangani.

Table 7.1: Salient features of reservoir and hydropower systems in Pangani river system (PBWO/IUCN, 2009).

		NyM	Hale	NPF
Reservoir				
Commissioning	Year	1968	1965	1995
Catchment area	km^2	12,100	42,200	42,200
Max. supply level	m.a.s.l	688.45	331.0	177.5
Min. supply level	m.a.s.l	679.62	329.8	176.0
Total reservoir storage capacity	Mm^3	1,140	0.14	1.4
Active storage capacity	Mm^3	600	0.13	0.8
Long term average inflow	Mm^3 yr^{-1}	1,100	730	730
Residence time	T	200 days	1.6 hrs	9.6 hrs
Power plant equipment				
Turbine type		Vertical Francis	Vertical Francis	Vertical Francis
Installed capacity	MW	2×4	2×10.5	2×34
Max. design discharge	m^3 s^{-1}	35	45	45
Min. design discharge	m^3 s^{-1}	9.8	8.5	9
Max. operating head	m	27	63	170
Min. operating head	m	21	62	168
Machine efficiency	%	87	76	93
Average annual energy	GWh yr^{-1}	35	93	341
Firm annual energy	GWh yr^{-1}	20	55	201

The operation policy at NyM reservoir is mainly aimed at generating firm energy at the three HEPs (Moges, 2003). The firm energy production is the amount of energy available for production or transmission which is guaranteed at 90% reliability (TANESCO, 2014). A discharge of 15 m^3 s^{-1} (39×10^6 m^3 month^{-1}) is therefore maintained as the minimum discharge to guarantee firm energy production at Hale and NPF power stations (Andersson et al., 2006).

The energy production, transmission and distribution is managed by Tanzania Electric Supply Company Limited (TANESCO), a state owned utility company. The Energy and Water Utilities Regulatory Authority (EWURA) determines the electricity tariffs. The electricity tariff comprises of 3 segments namely generation, transmission, and distribution and supply (EWURA, 2012). The value for hydropower production can be estimated from the tariff derived from the generation segment. The price of power (for monthly usage of less than 50 KWh) to domestic consumers is subsidized by government and represents the generation costs (FBD, 2003). In 2012, the price was estimated at 130 Tsh KWh^{-1} (80 US$ MWh^{-1}). A similar tariff is charged for bulk supply to Zanzibar Island in addition to transmission charges of TSh 35 KWh^{-1}. The energy price doubles for the monthly energy usage of more than 50 KWh units (160 US$ MWh^{-1}), since the demand is met by dispatching the more expensive thermal power plants in the TANESCO system (EWURA, 2012). The cost of bulk electricity purchases from independent power producers (IPPs) supplied by thermal sources is approximately US$ 166 MWh^{-1} (MEM, 2013). The bulk energy price is comparable with the higher electricity tariff for high energy users (160 US$ MWh^{-1}). In Tanzania,

Songas and Independent Power Tanzania Limited (IPTL) are the largest IPPs and supply 29% of energy supplied by TANESCO.

NyM reservoir is the only HEP with a storage reservoir while Hale and NPF operate as run-of-river systems. Through the construction of the NyM reservoir, the annually flooded area has reduced and vast areas are now (extensively) inhabited. Regulation of the flow by the NyM reservoir therefore prevents flooding of this area to protect the communities. The release from NyM reservoir is therefore limited to 25 m^3 s^{-1} for the river banks not to overflow at Kirua (PBWO/IUCN, 2007). The low lying Kirua swamp riparian ecosystem consumes part of the release from NyM reservoir through ground water recharge, transpiration and/or evaporation, of about 5 to 6 m^3 s^{-1} (PBWO/IUCN, 2007; Turpie et al., 2003; Andersson et al., 2006). The minimum operating level at NyM reservoir provides for a minimum surface area of the reservoir of 40 km^2. The surface area is considered sufficient to provide for environmental benefits mainly from fisheries to the local communities in NyM reservoir (Musharani, 2012).

The construction of NyM and NPF reservoirs has reduced the sediment flows into the Pangani Estuary, and affected the balance between tidal dynamics and morphology (Sotthewes, 2008). The sediment imbalance has resulted in erosion of river banks and the estuary bay, which influenced the tidal flow movement and hence the salt intrusion. However, the salt intrusion has been limited by the minimum (high) discharge released from the NPF HEP located 72 km upstream of the estuary. Sotthewes (2008), using a steady state salinity distribution model (Savenije, 2005), showed that the salinity profile reaches up to 5.5 km with a minimum discharge of 10 m^3 s^{-1}. At 5 m^3 s^{-1} salt intrusion would increase exponentially to 32 km upstream of the Pangani Estuary.

The Pangani estuary is also rich in mangrove resources that offer ecosystem services to local populations. There are at least 8 species of mangroves that cover an area of 1,750 ha (Turpie et al., 2003). The mangroves are mainly harvested for construction purposes and mostly exported to Zanzibar. Mangroves require both high and low flow conditions (de Lacerda et al., 2002; Alleman and Hester, 2011). The high flow is necessary for abscission of propagule and dispersal while the low flows are needed for propagule establishment and development. According to de Lacerda et al. (2002), high flows should not be sustained for longer periods since this may cause invasion of fresh water gycophytes that normally out-competes mangroves. Presently, there is no study to show how much of the high flows is required for mangroves growth nor how much of the mangroves in the Pangani estuary were affected by the construction of the NyM dam.

The major users of the Lower Pangani therefore are NyM HEP, Kirua swamp, irrigation and urban water users, Hale and NPF HEP and the Pangani estuary. The competition between demands of these users is analysed through a hydro-economic model (IHEM) that is described in the following section.

7.3 Materials and Methods

The IHEM is the central component of the multi-objective analysis. This model is fed by flows generated by the STREAM model, a fully distributed hydrological model that accounts for green and blue water flows from the Upper Pangani into the Lower Pangani river basin at NyM reservoir (Kiptala et al., 2014). The IHEM then optimises the water flows with the mostly blue water in the lower basin using the GAMS (General Algebraic Modelling System) programming language (GAMS, 2015). Objective functions or demand functions are developed for key water users in the basin. To reduce the number of objective functions, desirable levels of some objectives (mainly non-monetary) have been predetermined either through field investigation, by stakeholders and/or by expert knowledge. These 'secondary' objectives specify firm energy requirement, water supply for smallholder irrigation and urban water use. Other 'secondary' objectives include flood flow restrictions, environmental flows at a sensitive and large wetland located in the basin (Kirua swamp) and minimum environmental flow at the estuary. All objective functions are described in the following sections.

The hydropower, supplementary and full irrigation benefit functions, which are valued in monetary terms, are considered as primary objectives subject to the other predetermined constraints based on desired levels of 'secondary' objectives. The trade-offs between various objectives are then identified by removing each constraint and evaluating the gains or losses to other water users compared to a base case. The computations were conducted for a wet (2008), dry (2009) and an average year (2010). An overview of the methodological framework is presented in Fig. 7.2.

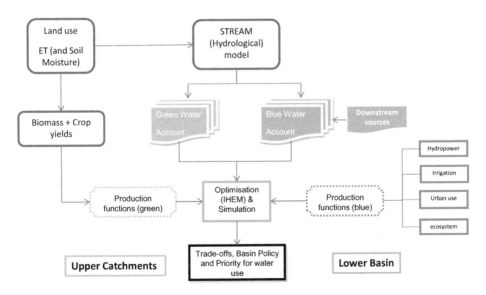

Fig. 7.2: Methodological framework for the multi-objective analysis.

7.3.1 STREAM hydrological model

A fully distributed hydrological model developed by Kiptala et al. (2014) is used to simulate stream flow for the period 2008-2010 for the Upper Pangani River Basin. The distributed model relies on remotely sensed data on actual total evaporation (Kiptala et al., 2013b) and land use and land cover (Kiptala et al., 2013a). The model is used to quantify green and blue water use such as supplementary and fully irrigated agriculture in the Upper Pangani upstream of the NyM reservoir and the blue water flow (river, groundwater) into the Lower Pangani hydro-system at the NyM reservoir (Fig. 7.3).

7.3.2 Hydro-Economic Modelling Approach

Whereas the formulation of an IHEM has no universal set-up, such a model adheres to the following essential requirements: a) consistent accounting of flows, water storages, and diversions, b) representation of demand for water and economic benefits for its use, c) network representation of a physical basin, and d) incorporation of institutional rules and policies (Cai et al., 2006). Water availability is determined by the water balance in the river system, while water demand is determined exogenously based on calculations of water requirements for irrigated agriculture, hydropower, issued water rights and estimated environmental flow requirements. Our model is schematized as a node-link network representing the spatial relation between various off- and in-stream demands in the river basin (Fig.7.3). The nodes represent the demand sites and links represent the river reaches. The nodes include simple nodes, source nodes at which inflows occur, reservoir nodes, and demand nodes. Each node should fulfill the water balance requirement.

For the source and simple node, there is no storage considered. The releases from these nodes are equal to the total inflows.

The equations that governs the mass balance for the source and simple nodes:

$$Q_{out}(n,t) = Q_{in}(n,t) \tag{7.1}$$

where

$$Q_{out}(n,t) = \sum_{j \in D_n} q_{n,j}(t) \tag{7.2}$$

$$Q_{in}(n,t) = \sum_{j \in U_n} q_{i,n}(t) \tag{7.3}$$

$q_{n,j}(t)$ represents flow from node n to node j, D_n is the set of all the nodes that are immediately downstream of node n and U_n is the set of all the nodes that are immediately upstream of node n.

$Q_{out}(n,t)$ is the release from the node n in period t, which is distributed over the downstream nodes. $Q_{in}(n,t)$ is the source of water or for the simple nodes the inflows at the time period t. Depending on the requirements at each node, water is diverted to users or remains in the river.

In the Kirua swamp (KS), the simple node that represents the release of flow to the area between the upstream node and downstream node is given by an empirical equation developed by IVO-NORPLAN (1997) (Eq. 7.4, all units in m^3 s^{-1}).

$$Q_{out}(KS,t) = -0.005 Q_{in}(KS,t)^2 + 0.9193 Q_{in}(KS,t) - 1.0308 \qquad (7.4)$$

Equation 7.4 holds for $Q_{in}(KS,t) < 25$ m^3 s^{-1} and excess inflows would drains out of the river system and is consumed by the wetlands.

Reservoir nodes are different as they consider storage. Here, we only consider the NyM storage reservoir and the following equation applies (Loucks et al., 1981) for period t:

$$(1 + \alpha_t)S(n,t) = (1 + \alpha_t)S(n,t-1) + Q_{in}(n,t) - Q_{out}(n,t) - \beta_t \qquad (7.5)$$

where

$$\alpha_t = \frac{A_a E_o(n,t)}{2} \qquad (7.6)$$

$$\beta_t = A_o E_o(n,t) \qquad (7.7)$$

$Q_{in}(n,t)$ and $Q_{out}(n,t)$ are defined in Eq. 7.2 & 7.3 and S is the storage (Mm^3), A_o is the water surface area corresponding to the dead storage volume (km^2), A_a is the water surface area per active storage volume above the dead storage level (km^2), and $E_o(n,t)$ is the evaporation rate in node n in period t (Loucks et al., 1981). The monthly evaporation rate is derived from pan evaporation measurements at NyM reservoir. The open water evaporation is computed using a pan coefficient factor of 0.81 (Kiptala et al., 2013b).

The water surface areas are computed from the reservoir area - volume equations (Eq. 7.8 & 7.9) derived from the original design report of NyM dam by Sir William Halcrow & Partners (1970):

$$V = (H/649.59)^{112.27} \qquad (7.8)$$

$$A = (H/651.59)^{88.15} \qquad (7.9)$$

where V is the reservoir volume (Mm^3), A is the surface area of reservoir (km^2) and H is the water level in metres above sea level (masl).

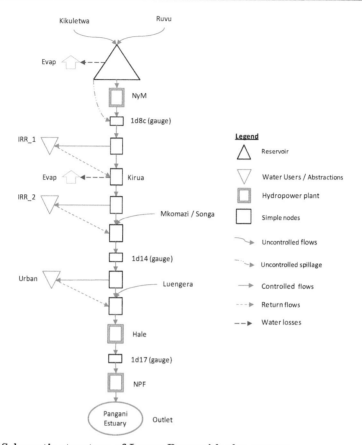

Fig. 7.3: Schematic structure of Lower Pangani hydro-system.

7.3.3 Multi-objective problem formulation for the Pangani hydro-system

Dynamic programming is a widely used optimization technique to determine optimal operating policies (Loucks et al., 1981). The objective of the reservoir operation problem optimization is to derive optimal release decisions as a function of variables describing the state of the system. The objective function therefore seeks to maximize benefits for each water sector subject to hydrological constraints (Eq. 7.1 – 7.7). For water use that comprise in-stream use such as hydropower, and off-stream functions such as irrigation, the water value is derived from the accumulated benefit functions to account for the cyclic nature of water use (Seyam et al., 2003; Pande et al., 2011).

The optimisation problem formulation (Eq. 7.10) is consistent with Tilmant et al. (2007), Kasprzyk et al. (2009) and Hurford and Harou (2014).

$$F(x) = \left(f_{ag\_I}, f_{ag\_R}, f_{hydro}, f_{firm}, f_{WR}, f_{estuary}, f_{Kirua} \right) \qquad (7.10)$$

$$\forall \; x \in \Omega$$

where x is the optimized water diversions and reservoir release for the set of water dependent sectors (Ω).

Two objective functions were considered that seek to maximize irrigated agriculture (f_{ag_I}) and rainfed agriculture (f_{ag_R}) in the upper catchments. Five objective functions were considered in the lower Pangani hydro-system. They include maximum hydropower (f_{hydro}), firm energy (f_{firm}), fulfilling water rights for smallholder farmers and urban water use in the mid-stream (f_{WR}), and environmental requirements for the Pangani estuary ($f_{estuary}$) and the Kirua swamp (f_{kirua}). The optimal requirements for f_{firm}, f_{WR}, $f_{estuary}$, f_{Kirua} are determined exogenously using field data and/or given by stakeholders. These water uses are considered as 'secondary' objectives that either used as constraints or whose deviations from their respective optimal requirement are minimized in the problem formulation (Eq. 7.10).

The multi-objective optimization would therefore seek to maximize three primary objectives: irrigated agriculture, rainfed agriculture and hydropower (Eq. 7.11) subject to secondary constraints (f_{firm}, f_{WR}, $f_{estuary}$, f_{Kirua}) and the hydrological constraints (Eq. 7.1 – 7.7).

$$F(x) = \left(f_{ag\_I}, f_{ag\_R}, f_{hydro} \right) \tag{7.11}$$

The model runs on a monthly time step with an optimisation period of 3 years (36 time steps). The problem formulation is solved using the GAMS MINOS solver (McKinney and Savitsky, 2003). The following sections detail the various water uses.

Fully irrigated agriculture

The objective function for the fully irrigated agriculture is to maximize the proceeds from the expansion of the sugarcane irrigation project to its potential.

$$\text{Maximize} \quad f_{ag\_I} = P_{n(s)} \times S_s \times Y_s \times \prod_t \left[\frac{W_r(t)}{W_d(t)} \right]^{l(t)} \tag{7.12}$$

$$S_s < S_{P(s)} \tag{7.13}$$

where $P_{n(s)}$ is the net gate price of sugar (US\$ kg^{-1}), S_s is the irrigation area (ha), $S_{p(s)}$ is the potential (sugarcane) irrigation area (ha), Y_s is the yield (kg ha^{-1}), t is the time index (month), $W_r(t)$ is the water diverted in each time period (Mm^3 month^{-1}), $W_d(t)$ water demand in each time period (Mm^3 month^{-1}) and $l(t)$ is the stress coefficient for sugar for each time step (equivalent to 1.2).

The potential irrigation area for irrigated sugarcane is 7,400ha, average sucrose yield of 10 tons ha^{-1} and net farm gate price of 0.25 US\$ kg^{-1} (Kiptala et al., 2016a).

Supplementary irrigated agriculture

The objective function for supplementary irrigated agriculture (in the upper catchment) is to enhance yields in rainfed systems by increasing productive transpiration (T) through supplementary irrigation. The impact of enhancing green water use would result in a reduction of blue water availability downstream (Kiptala et al., 2014). It was shown by Makurira et al. (2012) that an increase of productive T of up to 47% can be achieved in rainfed systems in the Pangani Basin. An increase in total ET (includes soil evaporation) of 30% can achieve relatively high T since part of soil evaporation would be shifted in favour of T. The concept of vapour shift in green water use has been described in more details by Rockström (2003). Kiptala et al. (2016a) developed an analytical relationship between biomass production (B_{acc}) and ET for rainfed and supplementary irrigated agriculture, Eq. 7.14 and Eq. 7.15 respectively.

$$B_{acc\ rainfed} = 3.9ET - 2.1 \tag{7.14}$$

$$B_{acc\ suppl\ irr} = 5.3ET - 2.6 \tag{7.15}$$

where B_{acc} is the accumulated biomass production in kg ha^{-1} yr^{-1} and ET is the total evapotranspiration in m^3 ha^{-1} yr^{-1}. The management option for enhanced green water use in rainfed systems involves both rainwater harvesting, and soil conservation and the use of fertilizers. An average rate of change in biomass production from Eq. 7.14 and Eq. 7.15 is adopted since the interventions would yield a hybrid agricultural system. Rainfed maize crop is considered with a potential area for improvement (high rainfall areas) of 36,000 ha.

The change in biomass production, B_{acc} (kg ha^{-1} yr^{-1}) due to an increase in green water use (Q_{g_b}) is converted into yield (Y_{mz}) using an effective harvest index of 0.35 for maize (i.e. $Y_{mz} = 0.35 \times B_{acc}$) (Wiegand et al., 1991; Kiptala et al., 2016a).

The objective function to be maximized therefore becomes Eq. 7.16:

$$\text{Maximize} \quad f_{ag\_R} = P_{n(mz)} \times Y_{mz} \times \sum_{t}^{T} \left(\frac{Q_{g\_b}}{Q_{g\_d}} \right) \tag{7.16}$$

$$Q_{g\_b} < \left(Q_{g\_d} \times S_{p(r)} \right) \tag{7.17}$$

where $P_{n(mz)}$ is the (net) farm gate price of maize (US\$ kg^{-1}), Y_{mz} is the additional yield per hectare for maize (kg ha^{-1}), Q_{g_b} is the additional green water use in rainfed area per month (Mm^3 month^{-1}), Q_{g_d} is the additional green water demand per hectare (Mm^3 ha^{-1} month^1) and $S_{p(r)}$ is the potential rainfed area (ha).

The reduction of soil evaporation (E_s) in supplementary irrigated crops (maize) in the Upper Pangani River Basin was also considered as an intervention. The reduction in E_s by 15% is considered feasible and results in a water saving (Q_{ws}) that has been quantified using the STREAM model (Kiptala et al., 2014).

Hydropower production

The production function of hydropower is used to derive its benefit function. The production function is a nonlinear function of the head (storage) and release variables. Power output P_y (Nm/s or W) and energy output E_y (Nm or Ws) is a function of discharge Q_p and head H_e derived using the following equations (Revelle, 1999).

$$P_y = e_t e_g \rho g Q_p H_e \tag{7.18}$$

where Q_p is the plant discharge (m^3 s^{-1}), ρ is the density of water (kg m^{-3}), g is the acceleration due to gravity (~9.81 m s^{-2}), H_e is the effective water head (m) (static water head – head loss) and $e_t e_g$ is the turbine and generator efficiency.

$$E_y = P_y \times T_g \qquad (7.19)$$

where, E_y is the Energy output (W s) and T_g is the generating time (s).

The optimization problem can be made linear by assuming that the production of hydroelectricity is dominated by the release term and not by the head (or storage) term. This assumption is valid as long as the difference between the maximum and minimum heads is small compared to the maximum head (Archibald et al., 1999; Wallace and Fleten, 2003). This assumption was used for main HEPS; Hale and NPF run-of-river systems where the difference between maximum and minimum operating head is small (Table 7.1).

The objective function for hydropower is to maximize the hydropower benefits from the water release at NyM reservoir. The bulk hydropower energy price is US\$ 80 MWh^{-1}. The opportunity cost for hydropower is estimated from cost of despatching alternative thermal sources or cost of bulk electricity purchases from IPPs which is equivalent to US\$ 160 MWh^{-1}. A similar approach was adopted by Kiptala et al. (2010) and Hurford and Harou (2014) for the Tana River Basin in Kenya.

$$\text{Maximize} \qquad f_{hydro} = \sum_{1}^{36} \sum_{i} (\text{Re }venue) \text{; } \in \{Nym, Hale, NPF\} \qquad (7.20)$$

Eq. 7.20 is subject to the minimum monthly firm energy requirement at the NyM, Hale and NPF hydropower stations (Table 7.2) formulated as secondary objective in Eq. 7.21.

$$\text{Minimize} \qquad f_{firm} = \sum_{1}^{36} \sum_{i} (Deficit \rightarrow Firm\_Energy) \text{; } \in \{Nym, Hale, NPF\} \qquad (7.21)$$

Irrigation and municipal water rights

According to the Pangani Basin office, there are 29 water abstraction canals for smallholder agriculture in the Lower Pangani Basin. The irrigation canals supply water to community development projects for food production, domestic use as well as for livestock. By 2010, water rights totalling 3.12 m^3 s^{-1} for water abstractions had been issued to various water user groups. Lemkuna, Naururu, Ngage are the main irrigation canals with a water right of 0.5 m^3 s^{-1} each, used mainly for rice and maize cultivation. An assessment of water flows between gauge station 1d8c at NyM and 1d14 at Korogwe (Fig. 7.1), less water uses/losses at Kirua using Eq. 7.4, showed a consistent irrigation water use of 3.14 m^3 s^{-1}. The municipal water use at Korogwe Township has a water right of 0.83 m^3 s^{-1}. The water is mainly used for domestic water supply and to a smaller extent by a sisal factory.

Since the objective of water allocation to these users is social rather than economic, the optimization problem is formulated to minimize deficits to their water rights provisions for irrigation (I) and municipal (M). A similar approach was adopted by Gurluk and Ward, 2009.

$$\text{Minimize } f_{WR} = \sum_{1}^{36} \sum_{i} (Deficit \rightarrow WR) \text{; } \in \{I, M\} \qquad (7.22)$$

Removing the objective function reveals the trade-offs with other economic activities in the river basin.

Environmental requirements

Demand curves for environmental benefits can be derived from the environmental goods and services that are provided to sustain ecosystems and to the environment by the water use. This however requires detailed environmental valuations that are linked to the hydrologic (supply) conditions and environmental benefits. This information is difficult to derive or estimate though there is a general understanding that valuing water should also account for environmental and social values (GWP, 2000; Hermans et al., 2006). An alternative approach is to remove the environmental flows from the objective function and treat them as additional constraints, thereby giving them priority (Gandolfi et al., 1997; George et al., 2011). The flow regime representing the lower bounds, i.e. the minimum flow requirements in space and time or flow constraints, could then be changed in order to establish the trade-off relationship. This approach requires an accurate hydrological assessment of the environmental flow requirement for the river basin.

Water use at the Kirua swamp is conditioned in the model using Eq. 7.4 for flows less than 25 m^3 s^{-1} to account for water use in the wetland. The maximum flow of 25 m^3 s^{-1} is imposed (on inflow) to prevent overtopping of river banks and flooding of areas currently occupied by local populations. These flow constraints will be removed in the model to assess trade-off with other water users.

Presently, there is no study on the environmental flow requirements for mangroves growth in the Pangani estuary before or even after the construction of the NyM dam. The model requirement for a maximum flow during the dry period or the high flows during wet seasons is therefore unknown and is not considered in this study. The study however uses maximum flow targets to assess implicitly the environmental flows. Since there is no evidence of any damage to the mangroves since the construction of the dam, conclusions will be drawn on the sustainability of the flow targets assessments. The targets are set based on 1) for the minimum flow requirement during the dry season on the fixed (or maximum) releases from two downstream hydropower dams provided through its high water demands and 2) from the peak flows during the wet season provided by the unregulated flows from Mkomazi and Luengera tributaries. For the minimum environmental flow (low flow, dry season) for the estuary, the study adopts the discharge of 10 m^3 s^{-1}, to minimize the impact on salt intrusion (Sotthewes, 2008).

$$\text{Minimize } f_i = \sum_1^{36} \sum_i (Deficit)_i \in \{Kirua, Estuary\} \tag{7.24}$$

The secondary objectives are summarized in Table 7.2.

Table 7.2. Secondary objectives considered in the Pangani hydro-system optimization model.

Secondary Objective	Model constraints
f_{firm}	Minimum discharge of 39 Mm^3 month^{-1} (15 m^3 s^{-1}) at Hale and NPF HEPs to guarantee firm energy.
f_{WR}	2.3 Mm^3 month^{-1} (0.83 m^3 s^{-1}) urban & 8.1 Mm^3 month^{-1} (3.12 m^3 s^{-1}) small-scale irrigation water rights
$f_{estuary}$	26.4 Mm^3 month^{-1} (10 m^3 s^{-1}) at the outlet
f_{kirua}	Release, Q$_{out}$ (KS,t) at Kirua conditioned by Eq. 7.4 for Q$_{in}$ (KS,t) $\leq$ 65.0 Mm^3 month^{-1}(25 m^3 s^{-1})

Problem Formulation

The problem formulation is carried out in two phases. The first phase involves the blue water use in the Lower Pangani hydro-system. In this phase, the current demand case (base scenario) is used to validate the IHEM model and represents the baseline water balance of the Lower Pangani hydro-system. The secondary objectives functions are considered as constraints in the base scenario. The secondary objectives are removed one by one in subsequent scenarios (Table 7.3).

Table 7.3: Problem formulations for blue water use in the Lower Pangani hydro-system.

Scenario	Primary objective	Secondary objectives	Remarks
1 (base)	f_{hydro}	f_{firm}, f_{WR}, f_{Kirua}, $f_{estuary}$	*ALL*
2	f_{hydro}	f_{WR}, f_{Kirua}, $f_{estuary}$	*No firm energy*
3	f_{hydro}	f_{firm}, f_{Kirua}, $f_{estuary}$	*No Water rights*
4	f_{hydro}	f_{firm}, f_{WR}, $f_{estuary}$	*No Kirua*
5	f_{hydro}	f_{firm}, f_{WR}, f_{Kirua}	*No Estuary*

Subsequently, the IHEM model is integrated with the green water use options in the upper catchments through their production functions (Table 7.4). The intervention for the reduction in soil evaporation (E_s) by 15% in supplementary irrigation (mixed crops) through water conservation (Q_{ws}) is also evaluated with all the objective functions (scenario B) and base demand case (scenario C). This is a water saving scenario (Q_{ws}) in the upper basin. The firm energy secondary objective is minimized and the other socio-environmental objectives (water rights and environment) are prioritized.

Table 7.4: Problem formulations for green and blue water use in Pangani Basin.

Scenario	Primary objectives	Secondary objectives	Remarks
A	f_{hydro}, f_{ag_I}, f_{ag_R}	*ALL*	*All objective functions*
B	f_{hydro}, f_{ag_I}, f_{ag_R}, Q_{ws}	*ALL*	*All objective functions plus water savings in agric.*
C	f_{hydro}, Q_{ws}	*ALL*	*Base scenario plus water saving in agric.*

7.4 RESULTS AND DISCUSSIONS

This section provides the results of the optimized scenarios, starting with the present demand (base scenario) that is used to validate the simulated results and generate the baseline water balance for the Lower Pangani hydro-system. The problem formulation scenarios are analyzed and discussed in subsequent sections.

7.4.1 Model validation

For scenario 1 (base), the goodness of fit between the observed and simulated water levels at NyM reservoir and the discharge at Korogwe (1d14) and the NPF (1d17) gauge stations are estimated using the coefficient of determination (R^2). The actual discharges at the outlet of NyM reservoir (1d8c) were not available for the period of analysis. The energy production is also compared to firm energy production and the average historical energy production. In addition, the water balance for the system was calculated and compared with values obtained from the literature.

Comparison observed vs simulated discharge and reservoir level

Fig. 7.4 shows observed and simulated water levels computed by base scenario (1) at NyM reservoir under Scenario 1. The convergence of observed and simulated reservoir water level occurred after 8 time steps. The simulated and observed water levels after convergence showed a good correlation (R^2=0.99).

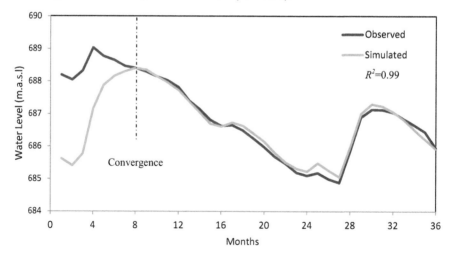

Fig. 7.4: Observed and simulated water levels in NyM reservoir for the period 2008 - 2010.

The simulated discharge and observed discharge at the downstream gauge stations 1d14 at Korogwe and 1d17 at Mnyuzi also showed reasonable correlations with R^2 of 0.7 and 0.8 respectively (Fig. 7.5).

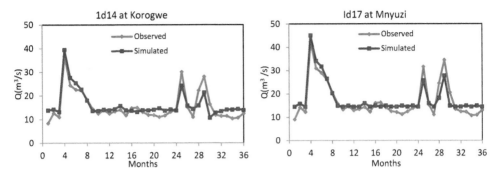

Fig. 7.5: Observed and simulated discharge at a) gauge 1d14 at Korogwe, b) gauge 1d17 at Mnyuzi for the period 2008 - 2010.

Station 1d17 is just downstream of Hale HEP where its simulated flow is influenced by the high flow requirement of 39 Mm^3 $month^{-1}$ (about 15 m^3 s^{-1}) needed to meet the firm energy requirements at the HEP. Both gauge stations located downstream of Kirua swamp. The lower performance of the model to simulate discharge compared to reservoir water level can partially be attributed to uncertainties in discharge measurements. Errors in estimating water losses from Kirua swamp and actual water abstractions especially during low flows may also have affected the performance, in particular during low flows

Comparison of observed versus simulated energy generation

The simulated and historical annual energy generation (firm, 5-yr and long-term) for each of the hydropower stations are provided and compared in Fig. 7.6. The long term historical energy production was available for the period 1985 - 2006 for NyM and Hale HEP and 1995 to 2006 for the NPF HEP (PBWO/IUCN, 2009). The 5-year historical hydropower was for the period 2002 - 2006 for all HEPs.

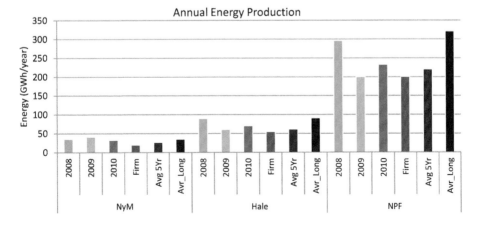

Fig. 7.6: Simulated annual energy production compared with firm, avg. 5 yr (2002 - 2006) and long term avg. for NyM, Hale and NPF Hydropower stations.

The simulated hydropower production is higher than the firm energy requirements for the HEPs, which is expected. The average long term hydropower is higher than both 5-yr and the simulated hydropower. This may be caused by declining water inflows into the Lower Pangani hydro-system due to increased water use by agriculture (PBWO/IUCN, 2007). The simulated hydropower for NyM reservoir shows small variance due to the regulated flow at NyM reservoir. High hydropower generation is realized in the dry year 2009 in NyM HEP due to increased outflow from NyM reservoir, an increase that is explained by the objective function of meeting the firm hydropower production by the large capacity HEPs downstream, and a subsequent lowering of the water level in NyM reservoir. In Hale and NPF HEPs, the hydropower production is higher in 2008 (wet year) due to higher (unregulated) discharge from Mkomazi and Luengera tributaries. The variability in energy production in 2008 (wet), 2009 (dry) and 2010 (average) years is also due to uncontrolled inflows from Mkomazi and Luengera tributaries. There is general consistency between the average hydropower production over the simulated period (2008 - 2010) with the 5-yr historical data (2002 - 2006). There is also consistency in the intra-seasonal trend in the hydropower production for the run-of-river Hale and NPF HEPs

Water balance for Lower Pangani hydro-system

The simulated evaporation losses at NyM reservoir were estimated at 7.9 m^3 s^{-1}, about 28% of the total inflow (27.8 m^3 s^{-1}) into NyM reservoir for the period 2008 - 2010. The simulated evaporation is within the upper limit of the ranges of 4 - 8 m^3 s^{-1} reported in the literature (Turpie et al., 2003; Andersson et al., 2006; PBWO/IUCN, 2009). The NyM reservoir releases an average of 20 m^3 s^{-1} of which an average of 4 m^3 s^{-1} was utilized for environmental functions in Kirua swamp, and another 4 m^3 s^{-1} for irrigation and municipal water use. Mkomazi and Luengera rivers injected an additional 6 m^3 s^{-1} into the Lower Pangani hydro-system, yielding an average total flow of 18 m^3 s^{-1} into the Pangani estuary (Fig. 7.7). Overall this shows that the model is able to simulate the system credibly and was therefore used for the optimisation scenarios.

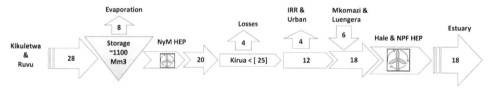

Fig. 7.7: The water balance (in m^3 s^{-1}) for Lower Pangani hydro-system for the period 2008 - 2010.

7.4.2 Problem formulation cases for Lower Pangani hydro-system

Table 7.5 presents the results of the five optimisation scenarios computed by the IHEM model based on five scenarios (Table 7.3). In scenario 1, the benefit functions of all water users are incorporated in the problem formulation (base scenario) and result in hydropower production of 355 GWh yr^{-1}, equivalent to US$ 28 million per year in energy revenue.

For the other scenarios where certain constraints are removed, hydropower production increases, this implies that maximising hydropower production affects other users in the basin. These trade-offs are further explained in the following section.

Table 7.5: Trade-off in hydropower between water users in Lower Pangani hydrosystem. Values in italics indicate years when the firm energy requirement is not met.

HEP	NyM (GWh/yr)			Hale (GWh/yr)			NPF (GWh/yr)			Total Energy	Revenue
Cases	2008	2009	2010	2008	2009	2010	2008	2009	2010	GWh/yr	US$ Million/yr
1 (base)	36	41	33	90	61	71	297	201	233	355	28
2	54	30	23	124	*47*	57	410	*156*	*188*	364	29
3	50	33	27	128	63	74	420	204	244	416	33
4	54	31	26	152	61	73	500	201	241	447	36
5	36	41	33	90	61	71	297	201	233	355	28
Firm	20			55			201			276	22

In scenario 2, the firm energy benefit function (f_{firm}) is removed from the objective function. The optimal operating policy maintains a lower head at the NyM reservoir to reduce evaporation losses through increased release to the downstream higher capacity HEP. The release policy provides more naturalized flow conditions where high flows are released during the wet year and low flows during dry years. The simulated evaporation loss at NyM reservoir is reduced from 7.9 to 6.7 $m^3 s^{-1}$. However, the water uses (losses) at Kirua swamp increase from 4.4 to 4.9 $m^3 s^{-1}$. There is a substantial increase in energy production of 165 GWh in year 2008 and a reduction of 69 GWh in both 2009 and 2010. The firm energy production is not maintained in 2009 and 2010 in Hale and NPF HEPs. In the end, an average annual energy production increase by 9 GWh (US$ 1 million) is realized. The firm energy requirement maintains moderate flows during drier years which tradeoffs with natural flow conditions required by the environment.

It is noteworthy that the cost for failing to meet the guaranteed firm energy production may be higher than the savings realized if emergency thermal systems, that have high short run marginal costs, are dispatched. However, if high capacity alternative energy sources like geothermal were available, then the hydropower production can be optimized within the naturalized flow regime. Examples of re-optimization techniques on reservoir operation or river restorations have been presented by Jacobson and Galat (2008). The re-designed reservoir policy will result in high energy production during wet seasons and low energy production during dry seasons. Alternatively, the bulk energy prices can be varied seasonally, as for the case in Kenya (Kiptala et al., 2010; Hurford and Harou, 2014), and independent providers can be invited to supply firm energy from thermal systems on long-term contracts. In such a case, the long term marginal cost of energy generation will be much lower.

Under scenario 3, the benefit function of maintaining water rights for smallholder irrigation and urban water use in the Lower Pangani is evaluated. Under this scenario, no water is diverted for these uses allowing more flow for HEP production at the hydropower stations downstream. The optimal operating policy would also maintain a lower water level at NyM due to reduced water rights requirements. The relatively small minimum flow requirements allow for flow conditions where higher flows are released during the wet seasons and lower flows during dry seasons. The evaporation at NyM reservoir would therefore reduce by 0.8 m^3 s^{-1} which is nearly balanced out by increased water uses (0.7 m^3 s^{-1}) at the Kirua swamp. In total, the energy production would increase by 14 and 47 GWh at Hale and NPF respectively mainly during the wet year (2008). There is no change in NyM HEP. The water right provision is equivalent to an average of US$ 5 million yr^{-1} in foregone hydropower benefits (smallholder agriculture (US$ 4 million yr^{-1}) and urban water use (US$ 1 million yr^{-1})). The water use by smallholder agriculture and urban is in competition with hydropower and also with environmental flows.

In scenario 4, the flow restrictions and water uses at the Kirua swamp of 4 m^3 s^{-1} are removed from the multi-objective function of the Pangani hydro-system. The optimal release policy at NyM reservoir also maintains a lower reservoir level to minimize evaporation losses due to reduced water requirements from Kirua swamp downstream. The simulated evaporation losses reduce by 1 m^3 s^{-1} at NyM reservoir to yield a total increased average outflow downstream of 5 m^3 s^{-1}. The annual energy production increased by 92 GWh in average over the period. The Kirua swamp therefore has an economic value of US$ 8 million yr^{-1} in foregone revenue to hydropower, using the bulk hydropower tariff of 80 US$ MWh^{-1}.

The objective function on the minimum environmental requirement of 10 m^3 s^{-1} at the Pangani estuary has no effect on the energy production (scenario 5). The high discharge requirement (15 m^3 s^{-1}) for firm energy at Hale and NPF HEP ensures that the minimum flow requirement for the estuary (downstream) is met and in many cases even exceeded. The uncontrolled inflows from the Mkomazi and Luengera tributaries also provide (and maintain) peak flows during the rainy seasons as observed in Fig. 7.5. Any future plans to control the river flows at Mkomazi and Luengera should be carefully considered taking into account high flow requirements for the mangroves and other ecosystem services at the estuary (de Lacerda et al., 2002; Alleman and Hester, 2011). For now, the level of high flow requirements for the estuary has not been explicitly established.

7.4.3 Problem formulation for green and blue water use

In this section, the three water management scenarios on increasing irrigated area, enhancing rainfed agriculture and reduced soil evaporation in the upper catchments were evaluated with the blue water uses in the Lower Pangani hydro-system. The STREAM model provided the hydrological input data into the IHEM model, considering all objective functions as used in scenario 1. Table 7.6 presents the optimization results.

Table 7.6: Green and blue water optimization scenarios in Pangani Basin. Values in italics indicate years when the firm energy requirement is not met.

HEP	NyM			Hale			NPF			Annual Totals		
										Energy	Revenue	Agric.
	2008	2009	2010	2008	2009	2010	2008	2009	2010	GWh	US$ Million	US$ Million
Base	36	41	33	90	61	71	297	201	233	355	28	-
A	27	31	25	76	*44*	59	249	*144*	*194*	283	23	55
B	28	32	26	79	*49*	62	261	*160*	205	301	24	55
C	44	41	33	104	61	71	342	201	233	376	30	-

In scenario A, the three multi-objective functions for hydropower, irrigated and rainfed agriculture are optimized. The optimization model diverts river flow to the total potential irrigation area (7,400 ha) for sugarcane and for 36,000 ha of rainfed maize (highland crop). The resulting minimum flow is 11 m^3 s^{-1}, below 15 m^3 s^{-1} that is required for firm energy. The water use (losses) at Kirua swamp reduces from 4.4 m^3 s^{-1} to 3.2 m^3 s^{-1}. The reduction (1.2 m^3 s^{-1}) represents about 27% of the additional requirements for both sugarcane irrigation and the rainfed system (4.5 m^3 s^{-1}). The average energy production reduces by 72 GWh yr^{-1}. The firm energy requirement for Hale and NPF is not met in dry (2009) and average years (2010). In terms of energy revenue, a total of US$ 5 million yr^{-1} is lost which represents an increased cost of US$ 10 million yr^{-1} if this power would have to be purchased from thermal sources. The revenue loss is much lower than the additional income to agriculture for sugarcane (US$ 19 Million yr^{-1}) and rainfed maize (US$ 36 Million yr^{-1}). The revenue for increased sugarcane production is calculated for a sucrose yield of 10 tons ha^{-1} (sucrose), farm gate price of 0.6 US$ kg^{-1} and a relative cost of production of 58% (Kiptala et al., 2016a). The area to be expanded for sugarcane irrigation is currently under grassland/woodlands which has a marginally low productivity, which is here neglected. The increase in transpiration in rainfed systems results in an increased biomass production of 15×10³ kg ha^{-1} yr^{-1} (produced in the two *Masika* and *Vuli* rainy seasons). The total revenue was derived from an effective harvest index of 0.35 and (net) farm gate price of 0.19 US$ kg^{-1}. It is noted that an additional 9,000 ha of rainfed agriculture can still be irrigated before the IHEM model is fully constrained by the minimum flow requirements of 10 m^3 s^{-1} at the estuary.

Fig. 7.8 shows the trade-off analysis among competing primary objectives (hydropower, irrigated sugarcane and supplementary irrigated maize) in the upstream and downstream catchments of the Pangani River Basin.

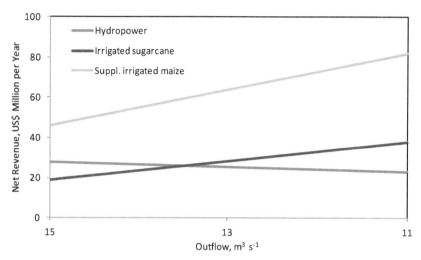

Fig. 7.8: Trade-offs between hydropower (downstream), agricultural water use (upstream) and the outflow (envir. flow) into the Pangani estuary, Scenario A.

In scenario B, the inflow into the river system is increased by the reduction of soil evaporation in the supplementary irrigated area (mixed crops) in the upper catchments. The additional inflow is 27% of the additional water requirements for agriculture. The additional inflow increases the energy production in the HEPs by 18 GWh yr⁻¹. The water uses at Kirua swamp slightly increase from scenario A, with a reduction of 23% to the base scenario. The minimum flow to the estuary also increases from 11 m³ s⁻¹ to 12 m³ s⁻¹.

In scenario C, the optimal policy maximizes hydropower production by maintaining a lower reservoir operating level. The lower reservoir operating level reduces evaporation losses by 0.4 m³ s⁻¹ that was balanced by increased water uses (losses) at Kirua swamp (due to the increased outflow). In total, the average energy production increased by 21 GWh yr⁻¹ (all during the wet year 2008) resulting in additional revenues of US\$ 2 million yr⁻¹ (savings of US\$ 4 million yr⁻¹ from thermal sources).

The increased agricultural water use in the upper catchments reduces benefits from hydropower, firm energy and the environment. The analysis shows that agricultural water use upstream of NyM reservoir has a higher marginal water value compared to hydropower. The marginal water value for the agriculture water use (blue water) is 0.35 US\$ m⁻³ for irrigated sugarcane and 0.37 US\$ m⁻³ for supplementary irrigated maize. This is much higher compared with the accumulated marginal water value of 0.05 US\$ m⁻³ for hydropower production (0.005 US\$ m³ (NyM) + 0.010 US\$ m³ (Hale) + 0.034 US\$ m³ (NPF), the water productivity being 0.05, 0.13 and 0.42 KWh m⁻³ for NyM, Hale and NPF respectively). This result is consistent with findings of multi objective optimization of water use between hydropower and irrigation in the Tana River Basin, Kenya (Kiptala et al., 2010; Hurford and Harou, 2014).

7.5 CONCLUSION

A multi-objective optimization model was developed and applied for the Pangani basin with a view to understand the trade-offs between all water uses, including green water use in the upper catchment. This has been a challenge since most if not all published optimization exercises only consider blue water uses. A hydrological model, STREAM, provided the hydrological information for green water flows in the upper catchments, which together with their economic productivity provided the production functions for the IHEM model. The distributed grid based STREAM model was linked to the IHEM (node-link) in such a way that all competing water users were integrated.

The model could therefore quantify the relationship between competing objectives involving blue water in the main stem of the river basin downstream NyM reservoir, and the green water uses in the run-off generating catchment upstream of the reservoir. The optimization analysis focused on three primary objective functions: i) hydropower production, ii) supplementary irrigated agriculture, and iii) fully irrigated agriculture. The analysis also considered five socio-environmental objectives, and a wet, a dry and an average year to represent the varying climate conditions in the basin.

The trade-off analyses show that hydropower, environment, urban and agriculture all have competing objectives. Firm energy that is guaranteed at 90% reliability maintains moderate flow conditions at all times, but competes with the environmental flow requirements (high and low flows). Neglecting the flow requirements for Kirua swamp resulted in an increase in hydropower production of US\$ 8 million yr^{-1}. At the estuary, the minimum environmental flow requirements were met as these were lower than the demand for firm energy. The high flow conditions, in this location, were provided and sustained by the uncontrolled inflow from Mkomazi and Luengera rivers. Future plans to control the river inflows from Mkomazi/Luengera tributaries should consider the environmental high flow requirement at the Pangani estuary, although this was not quantified.

Water saving by reducing soil evaporation losses in irrigated agriculture in the Upper Pangani resulted in increased hydropower revenue of US\$ 2 million yr^{-1}. This is equivalent to 33 US\$ ha^{-1} yr^{-1} of investment in soil and water conservation in irrigated agriculture, a potential for payment for environmental services (PES). The figures may double if the saving from expensive thermal energy sources were considered in the analysis. The increase water flow also enhanced environmental services in the lower catchment such as the observed enhanced water usage in the Kirua swamp.

As expected, increased water use for agriculture (rainfed maize and irrigated sugarcane) resulted in decreasing benefits for hydropower and firm energy. The estimated additional benefits for increased agricultural water use (US\$ 55 Million yr^{-1}) were much higher than the benefits foregone from hydropower (US\$ 5-10 Million yr^{-1}). However, the reduced flows also affect the downstream ecosystems, whose benefits were not quantified in this study. Furthermore, the study showed that improving rainfed maize through supplementary irrigation during rainy seasons has a slightly higher marginal water value than full scale sugarcane irrigation.

IHEM model provided the blue water balance for the Lower Pangani Basin and showed that the Upper Pangani River Basin contributes 82% of the total blue water.

Evaporation losses from the regulating reservoir constitute about 28% of the total in-flow into the reservoir. With these analyses, the decision makers in the Pangani River basin have the information required to decide whether to allocate additional water for upstream agricultural development or to trade-off with hydropower subject to the water requirements for ecosystem services. Since hydropower is a non-consumptive water user, the operating policy of the NyM reservoir could be further optimized for the conjunctive use with the environment. This may involve lowering the firm energy requirements which can be achieved by reducing dependency on hydropower in the river systems during dry periods. Alternative power sources, such as geothermal, and alternative institutional arrangements, e.g. through power purchase agreements, should be explored. However, this may result in higher energy prices during the dry seasons.

The study showed how this new methodological framework can be used by policy makers and stakeholders to identify holistically the impacts and opportunities of vari-ous water management decisions in the basin. The developed methodology may be useful for other closed river basins with green and blue water uses in the upper catchment and mainly blue water uses in the lower parts of a basin.

.

Chapter 8

CONCLUSIONS

Increasing competition over water resources in many river basins urge for increased water productivity of both green and blue water. Green and blue water follow distinct pathways and are associated with different water use practices. In African savannah's, rainfall-runoff processes are dominant in headwater catchments, while confined flows in the main river channels dominate downstream areas. Typically, natural land cover and rainfed agriculture dominate the landscape of headwater catchments, whereas irrigated agriculture, hydropower and environmental demands are the main water uses in the downstream part. Given the increasing demands for both green and blue water, the challenge is to build an integrated analytical system for the entire river basin, incorporating both upper and lower catchments processes. This would then allow for a quantitative assessment of upstream-downstream interdependences that can inform the establishment of an optimal management plan at the entire river basin scale. This thesis employed a number of approaches – some of which being innovations – to generate validated information for the optimal management of a heterogeneous, highly utilized and data scarce river basin in Africa. An accurate assessment of water availability, water use, water productivity and water value informed the identification of basin interdependencies, and allowed the formulation of a set of development options and their tradeoffs. Starting with hydrological modelling, the water balance of the basin has been verified at different spatial and temporal scales. Next, the complex green-blue water continuum has been quantified by modifying the structure of the hydrological model STREAM so that it can account for blue water contribution (irrigation and groundwater use) to the green water supplies. Actual evapotranspiration derived from RS data was a key input used to account for the use of blue water from groundwater and streams. Benefits from water use, in terms of water productivity, were assessed for both agricultural and natural landscapes. A key feature was the inclusion of ecosystem services in economic water productivity and their pitfalls in assessment. Lastly, the green and blue water uses in the upper catchment together with the predominantly blue water use in the lower part of the basin were integrated using an integrated hydro-economic model which can simulate water allocation decisions while accounting for ecosystem services and/or environmental flows. Although this analytical system has been developed for the Pangani River Basin in East Africa, it can be equally applied in river basins in the region and in other parts of the world, that have similar characteristics, i.e., extensive combined green – blue water use in the headwaters and nationally important natural and built infrastructure in the downstream part of the river basins.

8.1 ACADEMIC INNOVATION

8.1.1 Water balance assessment using RS data

A prerequisite for water resource management is the accurate assessment of water fluxes in a river basin. These include both blue water flows in rivers, wetlands and aquifers, as well as green water within the unsaturated zone. However, this presents a challenge in large, heterogeneous and data scarce river basins. The estimation of actual evapotranspiration, which is a major component of the water balance (in tropical climates), is not trivial, in particular in heterogeneous landscapes. This study employed different satellite measurements to establish the water fluxes in the Pangani basin in East Africa, in a spatially explicit manner.

Due to the high heterogeneity of land use in the Pangani basin, a detailed land use map was developed which yielded 16 land use classes that included classes relevant for hydrology and water management. The salient hydrological features are: wetlands, lakes and urban areas, natural land cover and modified landscapes fields. Different land use classes were distinguished using the phenological variability of vegetation, i.e. NDVI values at high spatial resolution, which was corroborated by expert knowledge of cropping calendar and ground observations. The resultant land use map managed to capture small scale informal smallholder irrigation developments that are prevalent in the river basin. The land use classification achieved acceptable accuracy levels for land use classification at a river basin scale and was consistent with the FAO-SYS framework for land suitability based on climate, topography and soil factors.

For each of these land use classes, actual evapotranspiration was computed on an 8-day timestep for three hydrological years (2008 to 2010). The time step (8-day) was chosen to correspond with agricultural water use processes and hydrological processes whose timescales are more than 8 days. An advanced interpolation technique was used to work around the cloudy pixels. The three hydrological years captured the varying weather conditions within the catchments. The actual evaporation estimates provided information on the intra-seasonal and inter-seasonal variability of actual evaporation for various land use types including the natural environment. The time series of actual evapotranspiration is a key boundary condition for hydrological modelling and water productivity analysis.

8.1.2 Modelling of green-blue water interaction and quantifying blue water use with a modified STREAM model

Current approaches to assess the impact of anthropogenic influences on water availability fail to account for a large part of the water use in rural Africa, in particular, spatially distributed water use such as small scale informal irrigation. In addition, estimates of this water use are often simplistic and notoriously inaccurate (assumed area, crop type diversity, actual irrigated area, application rate). This research therefore applied a novel approach of incorporating actual ET observed from satellites to calculate actual water abstraction at pixel scale. The new hydrological model used next to precipitation data, remotely sensed ET and soil moisture as input data into the model, decreasing the number of model variables for calibration - thus decreasing the problem of model equifinality widely associated with hydrological models. Furthermore, two hydrological landscapes (wetlands, hill-slope) that are typical in African landscapes were explicitly represented in the model configuration. Other hydrological processes such as capillary rise in natural landscapes were implicitly considered through the remotely sensed data. Comparing model results with observed river flow data validated the model; low flows during dry season were particularly well simulated, which is an important achievement when anthropogenic activities dominate river flows.

The modified STREAM model thus quantifies blue water use. The spatial nature of the analysis allowed for the assessment or tracking of blue water use at different nodes (either gauge station or dummy point) within the catchment. Also excessive and unregistered blue water users in the river systems could be identified.

8.1.3 Mapping ecological production and gross returns from water consumed in agricultural and natural landscapes

Water productivity is a key indicator for assessing water use efficiency for a given land use activity. Carbon credits and water yields (P-ET) provide insights into the water value society attaches to a certain cultural or natural land use activity. Standard Payment for Ecosystem Services (PES) touches also base with water yield, prevention of soil erosion and carbon sequestration, among others. It is therefore essential to evaluate the water productivity for both agricultural and natural landscapes in river basin scale. The methodological approach used combined various remote sensing models using Monteith's framework for ecological production. Instead of using default MODIS products, the biomass production and water yields were computed with the Surface Energy Balance Algorithm for Land (SEBAL) using locally calibrated model parameters. Grid biomass production (kg ha^{-1}) was estimated and converted into crop yield and amount of carbon sequestered. Gross returns were estimated using conversion factors for crop yield, carbon assimilates and market prices.

For the first time through this research, gross return from carbon credits and other ecosystem services were included in the concept of Economic Water Productivity (EWP). The analysis showed the level of EWP for various land use types and their scope for improvement. The EWP when formulated as production functions may be a basis for trade off analysis between green water and blue water uses if coupled to a hydrological model in a river basin. This could provide the basis for further sustained green growth in many African river basins, and demonstrate that spatially explicit information on EWP as estimated from earth observation is a vital piece in river basin planning and management.

8.1.4 Integrated hydro-economic modelling of green-blue water use

Through the integrated analysis of green and blue water use in a river basin, it has become clear that green and blue water use are hydrologically linked, and contributes significantly to economic productivity and wellbeing of local livelihoods in river basins. However, this is not always reflected in river basin planning and management, where national objectives (mainly blue water dominated uses such as hydropower production and irrigation) are sometimes prioritised over unregulated local uses of water, which typically includes large volumes of green water use. This research used a novel approached that linked the hydro-economics of green water use in the upper catchments of the Pangani Basin, with the blue water use further downstream in the main stem of the river. For the first time, green water use has been included in multi-objective optimization of water use using an integrated hydro-economic model (IHEM).

The study showed that agricultural water use in the form of supplementary and fully irrigated agriculture (upper catchment) achieves high levels of water productivity. It therefore competes with all the other water uses downstream, including hydropower. Supplementary irrigated agriculture (maize) had a slightly higher marginal value than fully irrigated sugarcane. Although the finding is specific to the Pangani basin, it re-emphases the high water value of using water resource (rainfall) at the starting point of the hydrological cycle (Pazvakawambwa and Van der Zaag, 2000; Hoekstra et al., 2001; Kijne et al., 2009; Bossio et al., 2011). For sustainability of the river basin, agricultural water uses need to be balanced with the environmental water requirements, that can be used conjunctively for hydropower production. However, the firm energy requirements that favours moderate low flow conditions over seasonal variability may have to be limited.

8.2 Uncertainty of RS data for Water Resource Planning

RS data was used to provide spatially explicit information on land use, actual evaporation and soil moisture and biomass production in the Upper Pangani River Basin (13,400 km^2). Data sources were mainly MODIS images at a resolution of 250 m and 8-day interval. The research validated the results with ground observations. These validations allow for a better insight in the uncertainty of remotely sensed data and their usefulness for water resource management.

The overall accuracy of the land use map of 85% was consistent with a recent review of previous studies that reported an average accuracy of 85% (STD 5%) (Karimi and Bastiaanssen, 2015). The accuracy levels for individual classes were generally over 70%. Low accuracies were observed for land use types with low spatial coverage, influenced by the moderate resolution of the satellite images. The Kappa coefficient factor that extents the accuracy limit by incorporating off-diagonal elements of the error matrices corrects this anomaly.

For the ET values estimated using SEBAL, the accuracy was inferred from other ET estimates since there were no *in-situ* measurements in the river basin. The ET estimates (250-m) for the entire river basin were found to be comparatively significant in variance but not with the mean at 95% confidence level to the global MODIS 16 ET (1 km resolution). The statistics between SEBAL ET and global MODIS 16 ET showed that the correlations (at monthly scale) were moderately fair (R=0.74; R^2=0.32; RMSE=34%; MAE=28%). At annual scale the correlations (R=0.91; R^2=0.70; RMSE=26%; MAE=24%) were better. The MAE range was within the 10-30% acceptable range of accuracy for comparative ET observations (also see Mu et al., 2011). The closing error of the annual basin water balance was 12% against the measured discharge (at the outlet). The bias (12%) was within the uncertainty (13%) level at 95% confidence interval for the P-ET estimates. The spatial distribution of SEBAL ET was consistent with the potential evaporation and also the temporal pattern of crop given by the crop coefficient factor (K_c).

The biomass production estimates relied on the evaporative fraction that is computed using SEBAL. The evaporative fraction was used to account for the spatio-temporal distribution of soil moisture in the estimation of light use efficiency. The uncertainty from the remotely sensed data (soil moisture) would therefore be within the range of SEBAL ET estimates. The biomass production was further subject to uncertainty from the model calibration parameters, i.e. the maximum light use efficiency and the effective harvest index that were calibrated against field data on crop yield. These parameters have experimentally verified ranges from the literature that limit the extent of errors in the validation process. It is observed that all the validation parameters for individual land use types used in this thesis were within these ranges. Moreover, the uncertainty of the biomass distribution to the mean was evaluated using the non-parametric bootstrapping technique. The uncertainty was low at below 1% at 95% confidence interval for all land use types.

A major limitation and challenge in all the RS computation of land use, ET and biomass is the persistent cloud cover especially for the land use types in the higher elevations around Mt. Kilimanjaro and Mt. Meru. In order to reduce the uncertainty associated with this limitation, the clouded pixels were removed and corrected using an advanced interpolation technique using neighbouring pixels or unclouded pixels from the previous or next available image. This procedure benefitted from the multi-temporal scales of the MODIS datasets and advanced computational capabilities of ERDAS software for which the level of uncertainty was limited. However, the uncertainty and the computational effort in the analysis can be eliminated by using radar based satellite data that are not affected by cloud cover (Bastiaanssen et al., 2012). Similarly, uncertainties that resulted from the use of moderate scale images on land use types of small coverage could be reduced by using RS data of finer resolutions such as from Landsat (30-m). The ET from Landsat can thereafter be up-scaled to MODIS (250-m) scales (e.g. in Hong et al., 2009). However, this will increase considerably the computational effort in the analysis of RS data in a river basin as the Upper Pangani (13,400 km^2).

8.3 RIVER BASIN MANAGEMENT IN THE PANGANI BASIN

The Pangani is a unique river basin with its high topographical range and associated range of climate and landscape characteristics. The basin is typically dominated by agriculture (mainly smallholder) in the upstream catchment and hydropower production in the downstream. The topological features have influenced the dominant hydrological processes in the landscape. Land use change through increasing agricultural activity (rainfed and irrigation) has changed the partitioning of green and blue water flows. This has significantly reduced the river flow downstream. The river basin has become water scarce and water related conflicts have emerged in the river basin (Sarmett et al., 2005; Komakech et al., 2011; Kiptala et al., 2013b).

For effective management of the water resources, one need to know how much water is available where and how it is being used. This knowledge is critical for water allocation decisions that the Pangani Basin Water Office has to make taking into consideration the principles of IWRM and sustainability. In a highly utilized and heterogeneous river basin, water availability and water use is highly varied. This therefore requires enormous investments in (the installation of) hydro-meteorological stations. Operation and maintenance of these hydro-met stations has proven a great challenge. For instance, out of 93 rain gauge stations installed over time in the river basin, less than 50% are presently operational. Some of the rain gauge stations at higher elevation (above the forest line of Mt. Kilimanjaro and Mt. Meru) that receive substantially high rainfall are no longer operational. There are also only six climate stations in the entire river basin. The climate data is essential for computing actual evapotranspiration, the principal water user in the river basin. The river basin therefore relies on few stream flow measuring devices to estimate its water uses. The Pangani River Basin is therefore a data scarce basin whose existing hydro-met measurements cannot be used effectively for water resource planning and management.

The water allocation and management by PBWO is mainly focussed on blue water (water in rivers, reservoirs, and wetlands as well as groundwater). The increasing uses of both green water and supplementary irrigation influence significantly the availability of blue water. This challenge is exacerbated by un-registered (illegal) water abstractions along the river canals and springs including excessive groundwater abstractions. Notable is the drying up of formerly perennial rivers in the dry seasons (Keller et al., 1998). The situation is made much more difficult by the presence of over 2000 traditional (furrow) irrigation canals that have a wide coverage with the majority been informal (Komakech et al., 2012). The irrigation canals consist of a complex and intricate network of furrows and ponds that presents big challenges for the estimation of water flows and their use. Commercial large scale agriculture (sugarcane) developed more recently has brought many technical and governance issues on water allocation especially in relation to the smallholder farmers. The usefulness of RS data to account for spatio-temporal water flows for all land uses is no longer in doubt in the Pangani basin.

The evaporative water use in the Upper Pangani River Basin was estimated at 94% of the total water resources (precipitation). The spatial P-ET estimates established the level of green water use for various land use types including the natural environment and provided their relative contribution and or effect to the downsatream hydrology. The IHEM model provided the extended blue water balance for the Lower Pangani Basin and showed that the Upper Pangani River Basin contributes 82% of the total blue water. The rest is provided by the downstream tributaries of Mkomazi and Luengera. There is thus a high degree of dependency of the lower catchment on the water resources from the upper catchment. The research also showed that the high flow requirement for Hale and NPF hydropower stations and the additional blue water flows from the downstream tributaries are currently sufficient to cover the minimum environmental flow requirement of the Pangani estuary. Investments in interventions to reduce soil evaporation in irrigated agriculture (upstream) may result in increased blue water flows downstream to the benefit of hydropower and the environment. Lowering of the firm energy production will result in a decrease in evaporation losses which is accompanied by seasonalized flows that can benefit the environment. The evaporation loss at NyM reservoir is presently high at 28% of the total blue water inflows into the reservoir. Further, the water uses at the Kirua swamp (20% of inflows) were found to be equivalent to US$ 8 million yr^{-1} of foregone hydropower production. These scenarios may create insights into the mutual dependencies between different water users in a river basin, which in turn may provide an opportunity to promote hydro-solidarity between the (upstream and downstream) users and between the various sectors. It also provides a framework for assessing tradeoffs in optimizing both green and blue water resources and the redistribution of benefits in the river basin.

The water balance analysis has shown that the river basin is closing. This means that any additional use of water may have direct implication for other uses including the environment. This requires a change in focus towards water use efficiency and ecological integrity. Currently, water allocation is based on water rights entitlement and political considerations. In this regard, the water productivity analysis provides a basis for re-allocation, opportunities for water saving and possible tradeoffs between water users. For instance, there is ample scope and opportunities for improving rainfed and supplementary irrigated agriculture which generates a higher economic return than the reduction in hydropower. Interestingly, up-scaling of supplementary irrigation was found to have a higher marginal water value (blue water) than full scale irrigation (large scale sugarcane). This may imply that up-scaling supplementary irrigated agriculture in the upper catchment of the Pangani river basin not only provides for rural livelihoods and significant economic returns but also has a large impact on downstream flows. The reduced inflows downstream thus not only result in lower hydropower but also reduced welfare for environmental and their associated vital ecosystem services. The relative economic importance of green and blue water as well as finding a balance with the environmental requirements should therefore not be ignored. Furthermore, the ecological productivity and specifically the gross returns from carbon credits and other ecosystem services can be formulated with an enhanced PES system and/or other conservation programmes for sustainable soil and water conservation programmes in the river system.

The PBWO can therefore develop with its stakeholders various optimal portfolios using the knowledge base and information and tools generated from this thesis to effectively manage the scarce water in the river basin. This may enhance water resource planning and management in an efficient and sustainable manner as well as provide for mechanisms for addressing the interdependence of the various stakeholders or water users in the river basin.

8.4 LESSONS FOR OTHER RIVER BASINS

The methodological approach presented in this thesis provides a comprehensive research framework to enhance water resource analysis at river basin scale. The methodology starts with scientifically derived and verified (validated) biophysical data. Since the river basin was expansive, grid based biophysical data were derived for the upper catchments that is dominated by green water use. Due to the size of the catchment (13,400 km^2), a moderate grid size satellite data was utilized at a temporal resolution of 8-day. The spatial hydrological link between green and blue water flows presented using the modified STREAM model highlighted the importance of landscape characteristics. The heterogeneity of the river basin was influenced by the high topographical range of over 5,000 m. In essence, an accurate land use and land cover map was found to be indispensible in this analysis. In the lower catchments, the vector based analysis of the biophysical data was used for the vast semi-arid plateau (30,000 km^2) dominated by blue water flows along the river channels. The methodological framework enabled a basin-wide analysis and integration of green and blue water uses for the entire river basin. Such an innovative approach (and methods) could be applied in any typical river basin ($10^2 - 10^5$ km^2) in the region.

The biophysical data on ET showed the general importance of green water in tropical climates. ET is normally a large component of the hydrological mass balance especially in highly utilized river basins. ET is therefore a key parameter in hydrological modelling. Recently developed surface energy balance methods that rely on satellite and/or airborne sensors have shown great potential in mapping ET. It is known that hydrological processes can be more realistically simulated at finer spatial and temporal resolutions and have to be locally validated. This therefore increases the computational effort and technical know-how on manipulation of RS data to deal especially with temporal challenges such as cloud cover. Nevertheless, the challenges of persistent cloud cover especially in the temperate climate could stimulate further innovate ways of using airborne sensors that are not affected by cloud cover. The high technical capacity required to generate ET data could also encourage research institutions to provide already processed information (at good levels of accuracy) through various public domain platforms for direct use by river basins around the world.

The spatial and temporal variability of water use presents challenges in water management between upstream and downstream water users. Validated hydrological models running various scenarios of water use for different configurations (or portfolios) can provide relevant information that can inform decision making. It was in this light that the modified STREAM model was developed to utilize RS validated ET data, given the high blue water use by informal supplementary irrigation. The development of this model presented a novel way of quantifying blue water use in a river basin. The model set-up could also improve hydrological and land surface energy simulations in landscapes dominated by blue water use. Such landscapes may include wetland ecosystems and landscapes with complex informal irrigation systems such as the Pangani. The hydrological model development using a flexible and open platform means therefore that the modified STREAM model can be applied in any other river basin.

The comprehensive economic water productivity assessment that included ecological services provided a framework for the sustainable management of natural ecosystems. Establishing the correct water value of natural ecosystem services would therefore stimulate conservation measures that could safeguard natural capital embodied in the environment that has hitherto often been associated with low water productivity.

REFERENCES

Abtew, W., 2001. Evaporation Estimation for Lake Okeechobee in South Florida. Journal of Irrigation and Drainage Engineering, 127, 140-147.

Abwoga, A.C., 2012. Modeling the impact of landuse change on river hydrology in Mara river basin, Kenya, Master's thesis, UNESCO-IHE Institute for Water Education, Delft.

Aerts, J.C.J.H., Kriek, M., Schepel, M., 1999. STREAM (Spatial tools for river basins and environment and analysis of management options): set up and requirements. Physics and Chemistry of the Earth, Part B: Hydrology, Oceans and Atmosphere, 24(6), 591-595.

Ahmed, M.D., Bastiaanssen, W.G.M., 2003. Retrieving soil moisture storage in the unsaturated zone from satellite imagery and bi-annual phreatic surface fluctuations. Irrigation Systems, 17(2), 3-18.

Ali, M.H., Hoque, M.R., Hassan, A.A., Khair, A.A., 2007. Effects of deficit irrigation on wheat yield, water productivity and economic returns. Agricultural water management, 92(3), 151 – 161, Doi:10.1016/j.agwat.2007.05.010.

Alleman, L.K., Hester, M.W., 2011. Reproductive ecology of black mangrove (Avicennia germinans) along the Louisiana coast: propagule production cycles, dispersal limitations, and establishment elevations. Estuaries and coasts, 34(5), 1068-1077.

Allen, R.G., Tasumi, M., Trezza, R., 2007. Satellite based energy balance for mapping evapotranspiration with internalized calibration (METRIC): Model. ASCE J. Irrigation Drainage Engineering, 133(4), 380-394.

Allen, R.G., Pereira, L.S., Raes, D., Smith, M., 1998. Crop Evapotranspiration - guidelines for computing crop water requirements, FAO Irrigation and Drainage Paper No. 56, FAO, Rome, Italy.

AMBIO, 2010. Scolel Te Programme Plan Vivo Annual Report 2009. San Cristobal de las Casas, Chiapas, Mexico.

Anderson, M. C., Kustas, W. P., Norman, J. M., Hain, C. R., Mecikalski, J. R., Schultz, L., Gonzalez-Dugo, M. P., Cammalleri, C., d'Urso, G., Pimstein, A., Gao, F., 2011. Mapping daily evapotranspiration at field to continental scales using geostationary and polar orbiting satellite imagery, Hydrology and Earth System Sciences, 15, 223–239, doi:10.5194/hess-15-223-2011.

Andersson, R., Wanseth, F., Cuellar, M., Von Mitzlaff, U., 2006. Pangani Falls Re-development Project in Tanzania. Sida Evaluation. Swedish International Development Cooperation Agency (SIDA), Stockholm, 67pp.

Ansar, A., Flyrbjerb, B., Budzier, A., Lunn, D., 2014. Should we build more large dams? The actual cost of hydropower megaproject development. BT centre for major programme management. Energy policy, 1 - 14. University of Oxford, UK.

Archibald, T.W., Buchanan, C.S., McKinnon, K.I.M., and Thomas, L.C., 1999. Nested benders decomposition and dynamic programming for reservoir optimization, Journal of the Operational Research Society, 50, 468–479.

Asrar, G., Myneni, R.B., Choudhury, B.J., 1992. Spatial heterogeneity in vegetation canopies and remote sensing of absorbed photosynthetically active radiation. Remote Sensing of Environment 41, 85–103.

Balsamo, G., Pappenberger, F., Dutra, E., Viterbo, P., van den Hurk, B., 2011. A revised land hydrology 5 in the ECMWF model: a step towards daily water flux prediction in a fully closed water cycle. Hydrological Processes, 25, 1046–1054.

Barker, R., Dawe, D., Inocencio, A., 2003. Economics of water productivity in managing water for agriculture, in Jacob Kijne et al. (Eds), water productivity in agriculture – limits and opportunities for improvements, comprehensive assessment of water management in agriculture, UK; CABI Publishing in association with International Water Management Institute, pp. 19-35.

Bashange, B.R., 2013. The spatial and temporal distribution of green and blue water resources under different landuse types in the Upper Pangani River Basin, Master's thesis, UNESCO-IHE Institute for Water Education, Delft.

Basnayabe, J., Jackson, P.A., Inman-Bamber, N.G., Lakshmanan, P., 2012. Sugarcane for water limited environments. Genetic variation in cane yield and sugar content in response to water stress. Journal of Experimental Botany, 63 (16), 6023–6033.

Bastiaanssen, W.G.M., 1998. Remote Sensing in Water Resources Management: the State of the Art. International Water Management Institute, Colombo, Sri Lanka.

Bastiaanssen, W.G.M., Ahmad, M.D., Chemin, Y., 2002. Satellite surveillance of evaporative depletion across the Indus Basin, Water Resource Research 38(12), 1273–1282, doi:10.1029/2001WR000386.

Bastiaanssen, W.G.M., Ali, S., 2003. A new crop yield forecasting model based on satellite measurements applied across the Indus Basin, Pakistan. Agriculture, Ecosystems and Environment 94, 321–340.

Bastiaanssen, W.G.M., Bandara, K.M.P.S., 2001. Evaporative depletion assessments for irrigated watersheds in Sri Lanka. Irrigation Sciences, 21, 1-15.

Bastiaanssen, W.G.M., Brito, R.A.L., Bos, M.G., Souza, R.A., Cavalcanti, E.B., Bakker, M.M., 2001. Low cost satellite data for monthly irrigation performance monitoring: benchmarks for Nilo Coelho, Brazil. Irrigation and Drainage Systems, 15, 53- 79.

Bastianssen, W.G.M, Karimi, P., Rebelo, L., Duan, Z., Senay, G., Muthuwatte, L., Smakhtin, V., 2014. Earth observation based assessment of the water production and water consumption of Nile Basin Agro-Ecosystems, Remote sensing, 6, 10306 – 10334; doi:10:3390/rs61110306.

Bastiaanssen, W.G.M., Menenti, M., Feddes, R.A., Holtslag, A.A.M., 1998a. A remote sensing Surface Energy Balance Algorithm for Land (SEBAL) 1. Formulation. Journal of Hydrology, 212-213, 198 - 212.

Bastiaanssen, W.G.M., Noordan, E.J.M., Pelgrum, H., Davids, G., Thoreson, B.P., Allen, R.G., 2005. SEBAL Model with Remotely Sensed Data to Improve Water Resources Management under Actual Field Conditions. Journal of Irrigation and Drainage Engineering, 131(1), 85-93.

Bastiaanssen, W.G.M., Pelgrum, H., Droogers, P., de Bruin, H.A.R., Menenti, M., 1997. Area-average estimates of evaporation, wetness indicators and top soil moisture during two golden days in EFEDA. Agriculture for Meteorology 87, 119–137.

Bastiaanssen, W.G.M., Pelgrum, H., Wang, J., Ma, Y., Moreno, J.F., Roerink, G.J., Van der Wal, T., 1998b. A remote sensing Surface Energy Balance Algorithm for Land (SEBAL) 2. Validation. Journal of Hydrology, 212-213, 213-229.

Bastiaanssen, W.G.M., Ahmad, M.D., Chemin, Y., 2002. Satellite surveillance of evaporative depletion across the Indus Basin. Water Resources Research, 38(12), 1273–1282. doi:10.1029/2001WR000386.

Bastiaanssen, W.G.M., Cheema, M.J.M., Immerzeel, W.W., Miltenburg, I.J., Pelgrum, H., 2012. Surface energy balance and actual evapotranspiration of the transboundary Indus Basin estimated from satellite measurements and the ETLOOK model, Water Resources Research, 48, W11512, doi:10.1029/2011WR010482.

Batjes, N.H., 2012. Projected changes in soil organic carbon stocks upon adoption of recommended soil and water conservation practices in the Upper Tana River catchment, Kenya. Land Degradation & Development, doi: 10.1002/ldr.2141.

Batra, N., Islam, S., Venturini, V., Bisht, G., Jiang, L., 2006. Estimation and comparison of evapotranspiration from MODIS and AVHRR sensors for clear sky days over the Southern Great Plains. Remote Sensing of Environment, 103, 1–15.

Bezuidenhout, C.N., Lecler, N.L., Gers, C., Lyne, P.W.L., 2006. Regional based estimates of water use for commercial sugar-cane in South Africa. Water SA 32(2), 219–222.

Bos, M.G., Burton, D.J., Molden, D.J., 2005. Irrigation and drainage performance assessment. Practical guidelines. CABI publishing, Cambridge, USA.

Boschetti, M., Bocchi, S., Stroppiana, D., Brivio, P.A., 2006. Estimation of parameters describing morpho-physiological features of Mediterranean rice varieties for modelling purposes. Italian Journal of Agrometeorology, 3: 40-49.

Boschetti M., Stroppiana D., Brivio P.A., Bocchi S., 2009. Multi-year monitoring of rice crop phenology through time series analysis of MODIS images. International Journal of Remote Sensing, 30 (18), 4643 - 4662.

Bossio, D., Jewitt, G., Van der Zaag, P., 2011. Editorial - Smallholder system innovation for integrated watershed management in Sub-Saharan Africa. Agricultural Water Management 98 (11), 1683-1686

Bouman, B.A.M., Humphreys, E., Tuong, T.P., Barker, R., 2006. Rice and water. Advances in Agronomy 92, 187 - 237.

Bouman, B.A.M., Lampayan, R.M., Tuong, T.P., 2007. Water management in rice: coping with water scarcity. Los Baños, (Philippines): International Rice Research Institute, 54 pp.

Brown, S.A.J., Gillespie, J.R., Lugo, A.E., 1989. Biomass estimation methods for tropical forests with application to forest inventory data. Forest Sciences 35(4), 881–902.

Burke, E. J., Shuttleworth, W.J., French, A.N., 2001. Using vegetation indices for soil-moisture retrievals from passive microwave radiometry, Hydrology and Earth Systems Sciences, 5(4), 671–678.

Burt, C.M., Howes, D.J., Mutziger, A., 2001. Evaporation Estimates for Irrigated Agriculture in California. ITRC Paper P 01-002. Irrigation Training and Research Center, San Luis Obispo, CA.

Cai, X., Ringler, C., Rosegrant, M., 2006. Modeling Water Resources Management at the Basin Level: Methodology and Application to the Maipo River Basin. Research Report, 149. International Food Policy Research Institute, Washington DC.

Campo, L., Caparrini, F., Castelli, F., 2006. Use of multi-platform, multi-temporal remote sensing data for calibration of a distributed hydrological model: an application in the Arno basin, Italy. Hydrological Processes 20, 2693-2712.

Campbell, J.B., 2002. Introduction to Remote Sensing. The Guilford Press, New York.

Carbon Market Watch, 2014. What is needed to fix the EU's carbon market? Recommendations for the market stability reserve and future ETs reform proposals. Carbon Market Policy Briefing, July 2014. European Union.

Carr, M.K.V., Knox J.W., 2011. The water relations and irrigation requirements of sugar cane (Saccharum officinarum): a review. Experimental Agriculture 47, 1-25.

Casanova, D., Epema, G.F., Goudriaan, J., 1998. Monitoring rice reflectance at field level for estimating biomass and LAI. Field Crops Research 55, 83–92.

Chang, M., and Lee, R.: Objective double mass analysis. Water resources research, 10(6), 1123-1126, 1974.

Cheema, M.J.M., Bastiaanssen, W.G.M., 2012. Local calibration of remotely sensed rainfall from TRMM satellite for different periods and spatial scales in the Indus Basin. International Journal of Remote Sensing, 33, 2603 – 2627..

Cheema, M.J.M., Bastiaanssen, W.G.M., 2010. Land use and land cover classification in the irrigated Indus Basin using growth phenology information from satellite data to support water management analysis. Agricultural Water Management, 97, 1541-1552.

Cheema, M.J.M., Immerzeel, W.W. Bastiaanssen, W.G.M., 2014. Spatial Quantification of Groundwater Abstraction in the Irrigated Indus Basin. Ground Water, 52: 25–36. doi: 10.1111/gwat.12027.

Cleugh, H.A., Leuning, R., Mu, Q., Running, S.W., 2007. Regional evaporation from flux tower and MODIS satellite data, Remote Sensing of Environment, 106, 285-304.

Congalton, R.G., Oderwald, R.G., Mead, R.A., 1983. "Assessing Landsat Classification Accuracy Using Discrete Multivariate Analysis Statistical Techniques", PERS, 49, 1671-1678.

Congalton, R.G., 1991. A review of assessing the accuracy of classifications of remotely sensed data. Remote Sensing of Environment, 37, 35 - 46.

Congalton, R.G., Green, K., 1999. Assessing the Accuracy of Remotely Sensed Data: Principles and Practices, Lewis Publishers.

Conover, W.J., 1980. Practical Nonparametric Statistics, Second edition. Wiley, New York.

Costanza, R., de Groot, R., Sutton, P., van der Ploeg, S., Anderson, S.J., Kubiezewski, I., Farber, S., Turner, R.K., 2014. Changes in the global value of ecosystem services. Global Environmental Change, 26, 152 – 158.

Costanza, R., d'Arge, R., de Groot, R., Farber, S., Grasso, M., Hannon, B., Limburg, K., Naeem, S., O'Neill, R.V., Paruelo, J., Raskin, R.G., Sutton, P., van den Belt, M., 1997. The value of the world's ecosystem services and nature capital. Nature, 387, 253-260.

Courault, D., Seguin, B., Olioso, A., 2005. Review on estimation of evapotranspiration from remote sensing data: from empirical to numerical modelling approaches. Irrigation and Drainage Systems, 19, 223-249.

Cuartas, L.A., Tomasella, J., Nobre, A.D., Nobre, C.A., Hodnett, M.G., Waterloo, M.J., de Oliveira, S.M., von Randow, R., Trancoso, R., Ferreira, M., 2012. Distributed hydrological modeling of a micro-scale rainforest watershed in Amazonia: Model evaluation and advances in calibration using the new HAND terrain model, Journal of Hydrology, Volumes 462–463, 10 15-27.

Dahmen, E.R., Hall, M.J., 1990. Screening of Hydrological Data. Tests for stationarity and Relative Consistency. Publication 49. International Institute for Land Reclamation and Improvement/ILRI, Wageningen, Netherlands, 58 pp.

De Bie, C.A.J.M., Khan M.R., Smakhtin, V.U., Venus, V., Weir, M.J.C., Smaling, E.M.A., 2011. Analysis of multi-temporal SPOT NDVI images for small-scale land-use mapping. International Journal of Remote Sensing, 32, 6673-6693.

De Groen, M.M. Savenije, H.H.G., 2006. A monthly interception equation based on the statistical characteristics of daily rainfall, Water Resour. Res., 42, 1–10, W12417, doi:10.1029/2006WR005013.

de Fraiture, C., Giordano, M., Liao, Y., 2008. Biofuels and implications for agricultural water use: blue impacts of green energy. Water Policy 10 (Suppl. 1), 67–81.

de Groot, R., Brander, L., van der Ploeg, S., Costanza, R., Bernard, F., Braat, L., Christie, M., Crossman, N., Ghermandi, A., Hein, L., Hussain, S., Kumar, P., McVittie, A., Portela, R., Rodriguez, L.C., ten Brink, P., van Beukering, P., 2012. Global estimates of the value of ecosystems and their services in monetary units. Ecosystem Services, 1, 50-61.

de Lacerda, L.D., Conde, J.E., Kjerfve, B., Alvarez-leon, R., Alarcon, C., Polania, J., 2002. Mangrove ecosystems: function and management: Springer, Niteroi, Brazil, pp 301.

de Troch, F.P., Troch, P.A., Su, Z., Lin, D.S., 1996. Application of Remote Sensing for Hydrological Modelling, in Distributed Hydrological Modelling, edited by M. B. Abbott and J. C. Refsgaard, pp. 165-192, Kluwer Academic Publishers, Dordrecht / Boston / London, 1996.

Dee, D.P., Uppala, S.M., Simmons, A.J., Berrisford, P., Poli, P., Kobayashi, S., et al., 2011. The ERA-Interim reanalysis: configuration and performance of the data assimilation system, Q. J. Roy. Meteorological Society, 137, 553–597.

Demarty, J., Chevallier, F., Friend, A.D., Viovy, N., Piao, S., Ciais, P., 2007. Assimilation of global MODIS leaf area index retrievals within a terrestrial biosphere model. Geophysical Research Letters, 34, 15, L15402.

Donald, C.M., Hamblin, J., 1976. The biological yield and harvest index of cereals as agronomic and plant breeding criteria. Advance Agronomy 28, 361–405.

Doorenbos, J., Pruitt, W.O., 1977. Crop water requirements. Irrigation and Drainage Paper No. 24, (revised) FAO, Rome, Italy. 144 p.

Efron, B., Tibshirani, R., 1993. An Introduction to the Bootstrap. Boca Raton, FL: Chapman & Hall/CRC.

Ehret, U., Zehe, E., 2011. Series distance - an intuitive metric to quantify hydrograph similarity in terms of occurence, amptitude and timing of hydrological events. Hydrology and Earth System Sciences, 15, 877-896.

Elsheikh, R., Shariff, A.R.B.M., Amiri, F., Noordin, Ahmad, N.B., Balasundram, S.K., Soom, M.A.M., 2013. Agriculture Land Suitability Evaluator (ALSE): A decision and planning support tool for tropical and subtropical crops. Computers and Electronics in Agriculture, 93, 98-110.

Engstrom, R., Hope, A., Kwon, H., Harazono, Y., Mano, M., Oechel, W., 2006. Modeling evapotranspiration in Arctic coastal plain ecosystems using a modified BIOME-BGC model. Journal of Geophysical Research, 111, G02021.

Enfors, E., Gordon, L., 2007. Analysing resilience in dryland agro-ecosystems: A case study of the Makanya catchment in Tanzania over the past 50 years. Land Degrad. Develop., 18, 680-696.

Enfors, E., Gordon, L., 2008. Dealing with drought: The challenge of using water system technologies to break dryland poverty traps. Global Environmental Change 18, 607–616.

ERDAS, 2010. ERDAS Field Guide. Leica Geosystems Geospatial Imaging, LLC. Norcross, GA 30092 - 2500, USA.

ERDAS, 2007. ERDAS Imagine Professional - Tour Guides. Leica Geosystems Geospatial Imaging, LLC. Norcross, GA 30092 - 2500, USA.

EWURA, 2012. Determination of multi-year cost reflective electricity tariff in Tanzania. available on www.ewura.go.tz.

Falkenmark, M., Folke, C., 2002. The ethics of socio-ecohydrological catchment management: towards hydrosolidarity. Hydrology and Earth System Sciences, 6(1), 1-9.

Falkenmark, M., Rockström, J., 2006. The new blue and green water paradigm: Breaking new ground for water resources planning and management, Editorial. ASCE, Journal of water resource planning and management, 132, 129 - 132.

Fanaian, S., S. Graas, Y. Jiang and P. van der Zaag, 2015. An ecological economic assessment of flow regimes in a hydropower dominated river basin: The case of the lower Zambezi River, Mozambique. Science of the Total Environment, 505, 464-473. [doi:10.1016/j.scitotenv.2014.10.033].

FAO, 1976. A framework for land evaluation. Food and Agriculture Organization of the United Nations, Soils Bulletin 32. FAO, Rome.

FAO, 2011. Crop calendar online database. http://www.fao.org/agriculture/seed/crop calendar (accessed on 7 July 2011).

FAO, 2004. Economic valuation of water resources in Agriculture. By Turner, K., Georgiou, S., Clark, R., Brouwer, R. FAO Water Reports, 24, pp 204.

FAO, 2013. FAOSTAT online database, available at link http://faostat.fao.org/. Accessed on December, 2013.

FAO, 1983. Guidelines: land evaluation for rainfed agriculture. Food and Agriculture Organization of the United Nations, Soils Bulletin 52. Rome, Italy.

FAO, 2007. Land Evaluation towards a revised framework. Food and Agriculture Organization of the United Nations, Rome, Italy.

FAO, 2011. Payments for ecosystem services and food security. ISBN 978-92-5-106796-3. Food and Agricultural Organization of the United Nations, Rome, Italy, 300pp.

FAO/IIASA/ISRIC/ISS-CAS/JRC, 2012. Harmonized World Soil Database (version 1.2). FAO, Rome, Italy and IIASA, Laxenburg, Austria.

Farah, H.O., Bastiaanssen, W.G.M., 2001. Spatial variations of surface parameters and related evaporation in the Lake Naivasha Basin estimated from remote sensing measurements. Hydrological Processes, 15, 1585-1607.

Farr, T., Rosen, P., Caro, E., Crippen, R., Duren, R., Hensley, S., Kobrick, M., Paller, M., Rodriguez, E., Roth, L., Seal, D., Shaffer, S., Shimada, J., Umland, J., Werner, M., Oskin, M., Burbank, D., Alsdorf, D., 2007. The Shuttle Radar Topography Mission, Rev. Geophysics, 45, RG2004, doi: 10.1029/2005RG000183.

Fereres, E., Orgaz, F., Gonzalez-Dugo, V., Testi, L., Villalobos, F. J., 2014. Balancing crop yield and water productivity tradeoffs in herbaceous and woody crops. Functional Plant Biology 41, 1009-1018.

FBD, 2003. Resource economic analysis of catchment forest reserves in Tanzania. United Republic of Tanzania, Ministry of Natural Resources and Tourism, Forestry and Beekepping Division, Dar es Salaam, 222pp.

Field, C.B., Randerson, J.T., Malmstrom, C.M., 1995. Global net primary production: combining ecology and remote sensing. Remote Sensing of Environment 51, 74–88.

Fisher, J.I., Mustard, J.F., 2007. Cross-scalar satellite phenology from ground, Landsat, and MODIS data. Remote Sensing of Environment, 109, 261-273.

Ford, C.R., Laseter, S.H., Swank, W.T., Vose, J.M., 2011. Can forest management be used to sustain water-based ecosystem services in the face of climate change? Ecological Applications, 21, 2049-2067. http://dx.doi.org/10.1890/10-2246.1.

Friedl, M.A., McIver, D.K., Hodges, J.C.F., Zhang, X.Y., Muchoney, D., Strahler, A.H., et al., 2002. Global land cover mapping from MODIS: Algorithms and early results, Remote Sensing of Environment, 83(1–2), 287-302.

Fu, B., Wang, J., Chen, L., Qiu, Y., 2003. The effects of land use on soil moisture variation in the Danangou catchment of the Loess Plateau, China. CATENA, 54 (1–2), 197-213.

GAMS, 2015. General Algebraic Modelling System (GAMS) and documentations [online] Available: http://www.gams.com.

Gandolfi, C., Guariso, G., Togni, D., 1997. Optimal flow allocation in the Zambezi river system, Water Resources Management, 11, 377–393.

George, B., Malano, H., Davidson, B., Hellegers, P., Bharati, L., Massuel, S., 2011. An integrated hydro-economic modelling framework to evaluate water allocation strategies II: Scenario assessment, Agricultural Water Management, 98 (5), 747-758.

Gerrits, A.M.J, 2005: Hydrological modelling of the Zambezi catchment from gravity measurements, Master's thesis, Delft, University of Technology, Delft, The Netherlands.

Gerrits, A.M.J., 2010. The role of inetrception in the hydrological cycle, PhD thesis, Delft, University of Technology, Delft, The Netherlands.

Giri, C., Jenkins, C., 2005. Land cover mapping of the greater Meso-America using MODIS data. Canadian Journal of Remote Sensing, 31, 274-282.

Gómez-Baggethun, E., Barton, D.N., 2013. Classifying and valuing ecosystem services for urban planning. Ecological Economics, 86, 235 - 245.

Gopalakrishnan, E., 2004. The role of dams in water and food security, health and poverty, hydropower & Dams. World Atlas, 2004, 11-12.

Gourbesville, P., 2008. Challenges for Integrated water resources management. Physics and Chemistry of the Earth, 33, 284-289.

Goutorbe, J.P., Lebel, T., Dolman, A.J., Gash, J.H.C., Kabat, P., Kim, Y.H., Monteny, B., Prince, S.D., Sticker, A., Tinga, A. and Wallace, J.S., 1997. An overview of HAPEX-Sahel: a study in climate and desertification. Journal of Hydrology, 188-189, 4-17.

Grimes, D.W., Wiley, P.L., Sheesley, W.R., 1992. Alfalfa Yield and Plant Water Relations with Variable Irrigation. Crop Science 32, 1381-1387.

Grossmann, M., 2008. Kilimanjaro Aquifer. In: Conceptualizing Cooperation for Africa's Transboundary Aquifer Systems, edited by: Scheumann, W. and Herrfahrdt-Pähle, E., DIE Studies Nr. 32, German Development Institute, 87-125, Bonn.

Gu, R.R., Li, Y., 2002. River temperature sensitivity to hydraulic and meteorological parameters. Journal of Environmental Management, 66, 43-56.

Gupta, H.V., Kling, H., Yilmaz, K.K., Martinez, G.F., 2009. Decomposition of the mean squared error and nse performance criteria: Implications for improving hydrological modelling, Journal of Hydrology, 377, 80–91.

Gürlük S., Ward, F.A., 2009. Integrated Basin Management: Water and Food Policy Options for Turkey. Ecological Economics, 68, 2666-2678.

GWP, 2000. Integrated Water Resources Management. TAC Background Paper 4. Global Water Partnership, Stockholm, Sweden, 71pp.

Haque, M.F.R., 2009. Validation of TRMM Rainfall data for hydrological applications in Pangani River Basin in Tanzania. MSc Thesis, WSE-HWR-09.05. UNESCO-IHE.

Hatfield, J.L., Asrar, G., Kanemasu, E.T., 1984. Intercepted photosynthetically active radiation estimated by spectral reflectance. Remote Sensing of Environment 14, 65–75.

Heinsch, F.A., Reeves, M., Votava, P., et al. 2003. User's guide GPP and NPP (MOD17A2/A3) products NASA MODIS land algorithm. NASA Goddard Space Flight Center, Greenbelt, MD, 57 p.

Hellegers, P.J.G.J., Soppe, R., Perry, C.J., Bastiaanssen, W.G.M., 2009. Combining remote sensing and economic analysis to support decisions that affect water productivity. Irrigation sciences 27, 243 -251.

Hemakumara, H.M., Chandrapala, L., Moene, A., 2003. Evapotranspiration fluxes over mixed vegetation areas measured from large aperture scintillometer. Agriculture Water Management, 58, 109-122.

Hermans, L.M., Hellegers, P., 2005. A "new economy" for water for food and ecosystem. Synthesis report for E-Forum results. FAO/Netherlands conference for food and ecosystems, 19pp.

Hermans, L.M., van Halsema, G.E., Mahoo, H.F., 2006. Building a mosaic of values to support local water resources management. Water Policy, 8, 415-434.

Hoekstra, A.Y., H.H.G. Savenije and A.K. Chapagain, 2001. An integrated approach towards assessing the value of water: A case study on the Zambezi basin. Integrated Assessment 2, 199-208.

Hong, S., Hendrickx, J.M.H., Borchers, B., 2009. Up-scaling of SEBAL derived evapotranspiration maps from Landsat (30 m) to MODIS (250m) scale. Journal of Hydrology, 370, 122-138.

Hong, Y., Hsu, K.-L., Moradkhani, H., Sorooshian, S., 2006. Uncertainty quantification of satellite precipitation estimation and Monte Carlo assessment of the error propagation into hydrologic response. Water Resources Research, 42, W08421, doi:08410.01029/02005WR004398.

Howell, T.A., Phene, C.J., Meek, D.W., Miller, R.J., 1983. Evaporation from screened class a pans in a semi-arid climate, Agricultural Meteorology, 29 (2), 111-124.

Hoy, R.D., Stephens, S.K., 1979. Field study of lake evaporation – analysis of data from phase 2 storages and summary of phase 1 and phase 2. Australian Water Resources Council Technical Report Paper No. 41.

Hsiao, T.C., Heng, L.K., Stedulo, P., Rojas-Lara, B., Raes, D., Fereres, E., 2009. AquaCrop - The FAO Model to simulate yield response to water. III. Parameterization and testing for maize. Agron Journal 101, 448 - 459.

Huffman, G.J., Adler, R.F., Morrissey, M.M., Bolvin, D.T., Curtis, S., Joyce, R., McGavock, B., Susskind, J., 2001. Global precipitation at one-degree daily resolution from multisatellite observations. Journal of Hydrometeorology 2, 36-50.

Hurford, A.P., Harou, J.J., 2014. Balancing ecosystem services with energy and food security assessing trade-offs for reservoir operation and irrigation investment in Kenya's Tana basin. Hydrology and Earth System Sciences Discussion, 11, 1343–1388.

Ibrom, A., Oltchem, A., June, T., Kreilein, H., Rakkibu, G., Ross, T., Panferov, O., Gravenhorst, G., 2008. Variation in photosynthetic light-use efficiency in a mountainous tropical rain forest in Indonesia. Tree physiology 28, 499 - 508.

Igbadun, H.E., Mahoo, H.F., Tarimo, A.K.P.R., Salim, B.A., 2006. Crop water productivity of an irrigated maize crop in Mkoji sub-catchment of the Great Ruaha River Basin, Tanzania, Agricultural Water Management, 85, 141-150.

ILRI, 2014. Livestock and fish market chains in Asia and Africa. International Livestock Research Institute. Available on www.ilri.org/node/1234. Accessed on 1/04/2014.

Immerzeel, W.M., Droogers, P., 2005. Calibration of a ditributed hydrological model based on satellite evapotranspiration. Journal of Hydrology, 349, 411-424.

Interagency Working Group on Social Cost of Carbon, 2009. "Social Cost of Carbon for Regulatory Impact Analysis under Executive Order 12866." US EPA Technical Support Document http://www .epa.gov/oms/climate/regulations/scc-tsd.pdf.

IUCN, 2003. The Pangani River Basin: A Situation Analysis, 1st edition. IUCN Eastern Africa Region Office, Nairobi. Scan house press limited. xvi + 104 pp.

IUCN, 2007. The Pangani River Basin: State of the Basin Report. IUCN Eastern Africa Regional Office, Nairobi.

IUCN, 2009. The Pangani River Basin: A Situation Analysis, 2nd edition. IUCN Eastern Africa Region Office, Nairobi. Scan house press limited. xii + 82 pp.

ISPRS, 2011. Global land cover mapping at finer resolution: Prof. Chen Jun. ISPRS e-Bulletin issue No. 5.

IVO-NORPLAN, 1997. Pangani Falls Redevelopment - Pangani River Training Project, Moshi, Tanzania.

Jacobson, R.B., Galat, D.L., 2008. Design of a naturalized flow regime – an example from the Lower Missouri River, USA. Ecohydrology, 1, 81–104.

Jarmain, C., Singels, A., Paraskevopoulos, A., Olivier, F., van der Laan, M., Taverna-Turisan, D., Dlamini, M., Münch, Z., Bastiaanssen, W., Annandale, J., Everson, C., Savage, M., Walker, S., 2014. Water use efficiency for selected irrigated crops determined with satellite imagery. Water Research Report No.TT 602/14. World Research Commission, Pretoria, South Africa.

Jensen, J.R., 1996. Introductory Digital Image Processing: A Remote Sensing Perspective. Englewood Cliffs, New Jersey: Prentice-Hall.

Jewitt, G., 2006. Integrating blue and green water flows for water resources management and planning. Physics and Chemistry of the Earth, 31, 753-762.

Jin,Y., Schaaf, C.B., Woodcock, C.E., Gao, F., Li, X., Strahler, A.H., et al., 2003. Consistency of MODIS surface BRDF/Albedo retrievals: 2. Validation. Journal of Geophysical Research, 108(D5), 4159.

Jonathan, M., Meirelles, M.S.P., Berroir, J.P., Herlin, I., 2006. Regional scale land use land cover classification using temporal series of MODIS data. ISPRS Commission VII, Mid-term Symposium "Remote Sensing: From Pixels to Processes", Enschede, the Netherlands, pp. 522-527.

Kalma, J.D., McVicar, T.R., McCabe, M.F., 2008. Estimating land surface evaporation: A review of methods using remotely sensed surface temperature data. Surveys in Geophysics, 29, 421-469.

Karimi, P., Bastiaanssen, W.G.M., 2015. Spatial evapotranspiration, rainfall and landuse data in water acccounting – Part 1: Review of the accuracy of the remote sensing data, Hydrology Earth System Sciences, 19, 533-532.

Karimi, P., Bastiaanssen, W.G.M., Molden, D., 2013a. Water Accounting Plus (WA+) – a water accounting procedure for complex river basins based on satellite measurements, Hydrology Earth System Sciences, 17, 2459-2472.

Karimi, P., Bastiaanssen, W.G.M., Molden, D., Cheema, M.J.M., 2013b. Basin-wide water accounting based on remote sensing data: an application for the Indus Basin. Hydrology Earth System Sciences, 17, 2473-2486.

Karimi, P., Bastiaanssen, W.G.M., Sood, A., Hoogeveen, J., Peiser, L., Bastidas-Obando, E., Dost, R.J., 2015. Spatial evapotranspiration, rainfall and landuse data in water accounting results for policy decisions in the Awash Basin, Hydrology Earth System Sciences, 19, 533-550.

Karssenberg, D., Burrough, P.A., Sluiter, R., de Jong, K., 2001. The PCRaster Software and Course Materials for Teaching Numerical Modelling in the Environmental Sciences. Transactions in GIS, 5(2), 99-110. [doi:10.1111/1467-9671.00070].

Kasprzyk, J.R., Reed, P.M., Kirsch, B.R., Characklis, G.W., 2009. Managing population and drought risks using many-objective water portfolio planning under uncertainty, Water Resources Research, 45, W12401, doi:10.1029/2009wr008121.

Keller, J., A. Keller, A., Davids, G., 1998. River basin development phases and implications of closure. Journal of Applied Irrigation Science, 33(2), 145-163.

Khan, M. S., Coulibaly, P., Dibike, Y., 2006. Uncertainty analysis of statistical downscaling methods, Journal of Hydrology, 319, 357–382.

Kijne, J., Barron, J., Hoff, H., Rockström, J., Karlberg, L., Gowing, J., Wani, S.P., Wichelns, D., 2009. Opportunities to increase water productivity in agriculture with special reference to Africa and South Asia: A report prepared by Stockholm Environment Institute, for the Swedish Ministry of Environment presentation at CSD 16, New York 14 May 2009. Stockholm: Stockholm Environment Institute, Sweden, 39pp.

Kilawe, E.C, Lusambo, L.P., Katima, J.H.Y., Augustino, S., Swalehe, N. O., Lyimo, B., 2001. Above ground biomass equations for determination of carbon storage in plantation forests in Kilombero district, Tanzania. International Forestry Review, 3(4), 317 - 321.

Kim, H.W., Hwang, K., Mu, Q., Lee, S.O., Choi, M., 2011. Validation of MODIS 16 Global Terrestrial Evapotranspiration Products in various climates and land cover types in Asia. KSCE Journal of Civil Engineering, 16, 229 - 238.

Kiptala, J.K., Tilmant, A., Van der Zaag P., 2010. Intersectoral water allocation in the Tana River Basin, Kenya. Water Resource Management, 82 - 106, 11th WaterNet/WARFSA/GWP-SA symposium, Victoria Falls, Zimbabwe.

Kiptala, J. K., Mohamed, Y., Mul, M., Cheema, M. J. M., and Van der Zaag, P., 2013a. Land use and land cover classification using phenological variability from MODIS vegetation in the Upper Pangani River Basin, Eastern Africa. Physics and Chemistry of the Earth, 66, 112-122.

Kiptala, J.K., Mohamed, Y., Mul, M.L., Van der Zaag, P., 2013b. Mapping evapotranspiration trends using MODIS images and SEBAL model in a data scarce and heterogeneous landscape in Eastern Africa. Water Resources Research, 49, 8495-8510, doi: 10.1002/2013WR014240.

Kiptala, J.K., Mul, M.L., Mohamed, Y., Van der Zaag, P., 2014. Modelling stream flow and quantifying blue water using modified STREAM model in the Upper Pangani River Basin, Eastern Africa, Hydrology and Earth System Sciences, 18, 2287-2303.

Kiptala, J.K., Mohamed, Y.A., Mul, M.L., Bastiaanssen, W.G.M., Van der Zaag, P., 2016a. Mapping ecological production and gross returns from water consumed in agricultural and natural landscapes. A case study of the Pangani River Basin, Tanzania. Submitted to *Water Research and Economics.*

Kiptala, J.K., Mul, M.L., Mohamed, Y.A., Van der Zaag, P., 2016b. Multi-objective trade-off analysis of the green-blue water uses in a highly utilized river basin in Africa. Submitted to *Journal of Water Resources Planning and Management, American Society of Civil Engineers (ASCE).*

Klein, I., Gessner, U., Kuenzer, C., 2012. Regional land cover mapping and change detection in Central Asia using MODIS time-series. Applied Geography, 35, 219-234.

Knight, J.F., Lunetta, R.L., Ediriwickrema, J., Khorra, S., 2006. Regional scale land use/land cover classification using MODIS-NDVI 250m multi-temporal imagery: a phenology based approach. GIScience and Remote Sensing, 43, 1-23.

Kollat, J.B., Reed, P.M., Maxwell, R.M., 2011. Many-objective groundwater monitoring network design using bias-aware ensemble Kalman filtering, evolutionary optimization, and visual 20 analytics, Water Resource Research, 47, W02529, doi:10.1029/2010wr009194.

Komakech, H.C., Condon, M., Van der Zaag, P., 2012. The role of statutory and local rules in allocating water between large- and small-scale irrigators in an African river catchment. Water SA, 38(1), 115-125.

Komakech, H., Van Koppen, B., Mahoo, H., Van der Zaag, P., 2011. Pangani River Basin over time and space: On the interface of local and basin level responses. Agriculture Water management, 98(11), 1740-1751.

Komakech, H.C., Van der Zaag, P., Van Koppen, B., 2012. The last will be first: Water transfers from agriculture to cities in the Pangani river basin, Tanzania. Water Alternatives, 5(3), 700-720.

Kongo, M.V., Jewitt, G.W.P., Lorentz, S.A., 2011. Evaporative water use of different land use in the Upper Thukela river basin assessed from satellite imagery. Agricultural Water Management, 98, 1727-1739.

Konrad, C.P, Warner, A., Higgins, J.V., 2012. Evaluating dam re-operation for freshwater conservation in the sustainable rivers project. River Research and Applications, 28, 777–792.

Korres, W., Reichenau, T.G., Schneider, K., 2013. Patterns and scaling properties of surface soil moisture in an agricultural landscape: An ecohydrological modeling study. Journal of Hydrology, 498, 89 – 102.

Kosgei, J.R., Jewitt, G.P.W., Kongo, V.M., Lorentz, S.A., 2007. The influence of tillage on field scale water fluxes and maize yields in semi-arid environments: A case study of Potshini catchment, South Africa. Physics and Chemistry of the Earth, 32 (15–18), 1117-1126.

Landis, R.J., Koch, G.G., 1977. The measurement of observer agreement for categorical data. Biometrics, 33 (1), 159 - 174.

Leblon, B., Guerif, M., Baret, F., 1991. The use of remotely sensed data in estimation of PAR use efficiency and biomass production of flooded rice. Remote Sensing of Environment 38, 147–158.

Lenhart, T., Eckhardt, K., Fohrer, N., Frede, H.-G., 2002. Comparison of two different approaches of sensitivity analysis. Physics and chemistry of the Earth, 22, 645-654.

Lehmann, E.L., 1975. Nonparametrics, Statistical Methods Based on Racks. Holden-Day, Inc, California.

Levene, H., 1960. Contributions to Probability and Statistics. Stanford University Press.

Li, A., Bian, J., Lei, G., Huang, C., 2012. Estimating the maximal light use efficiency for different vegetation through the CASA Model combined with Time-series remote sensing data and ground measurements. Remote Sensing 4, 3857-3876.

Liang, S.L., 2001. Narrowband to broadband conversions of land surface albedo I Algorithms. Remote Sensing of Environment, 76, 213–238.

Lillesand, T.M., Kiefer, R.W., 1994. Remote Sensing and Image Interpretation. New York: John Wiley & Sons.

Liu J., Williams J.R., Zehnder A.J.B., Yang H., 2007. GEPIC – modelling wheat yield and crop water productivity with high resolution on a global scale. Agricultural Systems, 94 (2), 478-493, doi: 10.1016/j.agsy.2006.11.019.

Liu, Y.B., De Smedt, F., 2004. WetSpa Extension. A GIS-Based Hydrologic Model for Flood Prediction and Watershed Management. Documentation and User

Manual. Department of Hydrology and Hydraulic Engineering. Vrije Universiteit, Brussel, Belgium, 2004.

LMC International, 2010. Worldwide Survey of Sugar and HFCS Production Cost (2000 - 2009). Overseas Development Institute, London, UK.

Long, D., Singh, V.P., 2012. A modified surface energy balance algorithm for land (M-SEBAL) based on a trapezoidal framework. Water Resources Research, 48, 1-24.

Loucks, D., Stedinger, J., Haith, D., 1981. Water resources systems planning and analysis. Prentice-Hall, NJ, USA, 559pp.

Lucht, W., Schaaf, C.B., Strahler, A.H., 2000. An Algorithm for the retrieval of albedo from space using semi-empirical BRDF models, IEEE Transactions. Geoscience and Remote Sensing, 38, 977-998.

Maas, S.J., 1988. Use of remotely sensed information in agricultural crop growth models. Ecological Modelling 41, 247–268.

Maidment, D.R., (Ed.) 1993. Handbook of hydrology New York: McGraw Hill.

Mainuddin, M., Kirby, M., 2009. Spatial and temporal trends of water productivity in the lower Mekong River Basin. Agricultural Water Management, 96, 1567 - 1578.

Makurira, H., Savenije, H.H.G., Uhlenbrook, S., 2010. Modelling field scale water partitioning using on-site observations in sub-Saharan rainfed agriculture. Hydrology and Earth System Sciences, 14, 627-638.

Malley, Z.J.U., Taeb, M., Matsumoto, T., Takeya, H., 2007. Environmental change and vulnerability in the Usangu plain, southwestern Tanzania: implications for sustainable development. International Journal of Sustainable Development and World Ecology, 14, 145–159.

Mbonile, M.J., 2005. Migration and intensification of water conflicts in the Pangani Basin, Tanzania. Habitat International, 29 (1), 41-67.

McCabe, M., Wood, E., 2006. Scale influences on the remote estimation of evapotranspiration using multiple satellite sensors. Remote Sensing of Environment, 105, 4, 271–285.

McDonnell, J.J., Sivapalan, M., Vache, K., Dunn, S., Grant, G., Haggerty, R., Hinz, C., Hooper, R., Kirchner, J., Roderick, M.L., Selker, J., Weiler, M., 2007. Moving beyond heterogeneity and process complexity: A new vision for watershed hydrology, Water Resources Research, 43, W07301, doi:10.1029/2006WR005467.

McKinney, D.C., Savitsky, A.G., 2003. Basic optimization models for water and energy management. Revision No. 6 available at www.gams.com, 178pp.

McMahon, T.A., Peel, M.C., Lowe, L., Srikanthan, R., McVicar, T.R., 2013. Estimating actual, potential, reference crop and pan evaporation using standard meteorological data: a pragmatic synthesis. Hydrology Earth System Sciences, 17, 1331–1363.

McVicar, T.R., Zhang, G., Bradford, A.S., Wang, H., Dawes, W.R., Zhang, L., Li, L., 2002. Monitoring regional agricultural water use efficiency for Hebei province on the North China Plain. Australian Journal of Agricultural Research, 53, 55-76.

MEM, 2013. Power system master plan. Ministry of Energy and Minerals. United Republic of Tanzania, 157pp.

Millennium Ecosystem Assessment, 2005. Ecosystems and Human Well-being: Current States and Trends. Island Press, Washington D.C., USA.

Minitab Inc, 2003. MINITAB Statistical Software, Release 14 for Windows. State College, Pennsylvania.

Misana, S., Sokoni, C., Mbonile, M., 2012. Land-use/cover changes and their drivers on the slopes of Mount Kilimanjaro, Tanzania. Journal of Geography and Regional Planning, 5(6), 151-164.

Miralles, D.G., Holmes, T.R.H., De Jeu, R.A.M., Gash, J.H., Meesters, A.G.C.A., Dolman, A.J., 2011. Global land-surface evaporation estimated from satellite-based observations. Hydrology Earth System Sciences, 15, 453–469.

Mkhwanazi, M., Chávez, J.L., Rambikur, E.H., 2012. Comparison of large apenture scintillometer and satellite - based energy balance models in sensible heat flux and crop evapotranspiration determination. International Journal of Remote Sensing Applications, 2(1), 24 - 30.

Mobbs, D.C., Cannell, M.G.R., Crout, N.M.J., Lawson, G.J., Friend, A.D., Arah, J., 1997. Complementarity of light and water use in tropical agroforests. 1. Theoretical model outline, performance and sensitivity. Forest Ecology and Management 102, 259 - 274.

Moges, S., 2003. Development and application of Hydrological Decision Support tools for Pangani River Basin in Tanzania. PhD Thesis, University of Dar es Salaam, Dar es Salaam University Press.

Mohamed, Y.A., Bastiaanssen, W.G.M., Savenije, H.H.G., 2004. Spatial variability of evaporation and moisture storage in the swamps of the upper Nile studied by remote sensing techniques. Journal of Hydrology, 289, 145-164.

Molden, D., Bekele Awulachew, S., Conniff, K., Rebelo, L. M., Mohamed, Y., Peden, D., Kinyangi, J., van Breugel, P., Mukherji, A., Cascão, A., Notenbaert, A., Demise, S.S., Neguid, M.A., Naggar G., 2009. Nile Basin Focal Project. Synthesis Report, Project Number 59. Challenge Program on Water and Food and International Water Management Institute, Colombo, Sri Lanka. x+149 pp.

Molden, D., Oweis, T.Y., Pasquale, S., Kijne, J.W., Hanjra, M.A., Bindraban, P.S., Bouman, B.A.M., Cook, S., Erenstein, O., Farahani, H., Hachum, A., Hoogeveen, J., Mahoo, H., Nangia, V., Peden, D., Sikka, A., Silva, P., Turral, H., Upadhyaya, A., Zwart, S., 2007. Pathways for increasing agricultural water productivity. In: Molden, D. (Ed.), Water for Food, Water for Life: A

Comprehensive Assessment of Water Management in Agriculture. Earthscan/IWMI, London, UK/Colombo, Sri Lanka, pp. 279–310.

Molden, D., Oweis, T.Y., Steduto, P., Bindraban, P., Hanjra, M.A., Kijne, J.W., 2010. Improving agricultural water productivity: between optimism and caution. Agricultural Water Management 97, 528–535.

Molden, D., Sakthivadivel, R., 1999. Water Accounting to Assess Use and Productivity of Water, International Journal of Water Resources Development, 15:1-2, 55-71, doi: 10.1080/07900629948934.

Molden, D., Sakthivadivel, R., Samad, M., Burton, M., 2005. "Phases of River Basin Development: The Need for Adaptive Institutions." In M. Svendsen, ed., Irrigation and River Basin Management: Options for Governance and Institutions. Wallingford, UK: CABI Publishing.

Molle, F., Wester, P., 2009. River basin trajectories: an inquiry into changing waterscapes. In: Molle, F. and Wester, P. eds. River basin trajectories: societies, environments and development. CAB International, Wallingford Oxfordshire, United Kingdom.

Molle, F., Wester, P., Hirsch, P., 2005. Water for food, water for life. River basin development and management. International Water Management Institute (IWMI), Earthscan, London, NW1 0JH, UK, 16 (4), 585 - 624.

Moncrieff, J., Monteny, B., Verhoef, A., Friborg, T., Elbers, J., Kabat, P., de Bruin, H., Soegaard, H., Jarvis, P., Taupin, J., 1997. Spatial and temporal variations in net carbon flux during HAPEX-Sahel. Journal of Hydrology, 188-189, 563-588.

Monfreda, C., Wackernagel, M., Deumling, D., 2004. Establishing national capital accounts based on detailed Ecological Footprint and biological capacity assessments. Land Use Policy, 21, 231 – 246.

Monteith, J.L., 1965. Evaporation and environment. In: B.D. Fogg, (Ed.), The State and Movement of Water in Living Organism, Symposium of the society of experimental biology, 19, 205–234.

Monteith, J.L., 1972. Solar radiation and productivity in tropical ecosystems. Journal applied ecology 9, 747 - 766.

Moran, M.S., Maas, S.J., Pinter, P.J., 1995. Combining remote sensing and modeling for estimating surface evaporation and biomass production. Remote Sensing Reviews 12, 335–353.

Morse, A., Tasumi, M., Allen, R.G., Kramber, J.W., 2000. Application of the SEBAL methodology for estimating consumptive use of water and stream flow depletion in the Bear River Basin of Idalo through remote sensing. Earth Observation System Data and Information System Project Report. The Raytheon Systems Company. Idalo, USA.

Mostert, E., van Beek, E., Bouman, N.W.M., Hey, E., Savenije, H.H.G., Thissen, W.A.H., 1999. River basin management and planning, Proceedings of the

International Workshop on River Basin Management, IHP-V, Technical Documents in Hydrology, No. 31, The Hague, 27–29 October 1999, 24 – 55pp.

MOWI, 2009. Irrigation Database, Kilimanjaro and Arusha Regions. Dar es Salaam: Ministry of Water and Irrigation, Government of the United Republic of Tanzania.

Mujwahuzi, M.R., 2001. Water use conflicts in the Pangani basin. In: Ngana, J.O. (Ed.), Water Resources Management in the Pangani River Basin; Challenges and Opportunities. Dar es Salaam University Press, Dar es Salaam.

Mul, M.L., 2009. Understanding Hydrological Processes in Ungauged catchment in Sub-Saharan Africa. PhD dissertation. TU Delft/UNESCO-IHE, Delft, the Netherlands.

Mu, Q., Heinsch, F.A., Zhao, M., Running, S.W., 2007. Development of a global evapotranspiration algorithm based on MODIS and global meteorology data, Remote Sensing of Environment, 111, 519-536.

Mu, Q., Zhao, M., Running, S.W., 2011. Improvements to a MODIS Global Terrestrial Evapotranspiration Algorithm. Remote Sensing of Environment, 115, 1781 - 1800.

Mujwahuzi, M.R., 2001. Water use conflicts in the Pangani basin. In: Ngana, J.O. (Ed.), Water Resources Management in the Pangani River Basin; Challenges and Opportunities. Dar es Salaam University Press, Dar es Salaam.

Murphy, A. H., 1995. The coefficients of correlation and determination as measures of performance in forecast verification. Weather and forecasting 10(4): 681-688.

Musharani, E., 2012. Local benefits and burdens for Nyumba ya Mungu reservoir. MSc. Thesis, UNESCO-IHE, Delft, 72pp.

Muthuwatta, L.P., Ahmad, M., 2010. Assessment of water availability and consumption in the Karkleh River Basin, Iran - Using remote sensing and geo-statistics. Water Resources Management, 24, 459 - 484.

Myneni, R.B., Hoffman, S., Knyazikhin, Y., Privette, J.L., Glassy, J., Tian, Y., et al., 2002. Global products of vegetation leaf area and fraction absorbed PAR from year one of MODIS data, Remote Sensing of Environment, 83(1–2), 214-231.

Nagler, P.L., Scott, R.L., Westenburg, C., Cleverly, J.R., Glenn, E.P., Huete, A.R., 2005. Evapotranspiration on western U.S. rivers estimated using the Enhanced Vegetation Index from MODIS and data from eddy covariance and Bowen ratio flux towers, Remote Sensing of Environment, 97(3), 337–351.

Namayanga, L.N., 2002. Estimating terrestrial carbon sequestered in above ground woody biomass from remotely sensed data. SEBAL and CASA algorithms in a semi-arid area of Serowe - Botswana. Msc Thesis in Environmental Systems analysis and management. ITC, Enschede, the Netherlands.

Nash, J.E., Sutcliffe, J.V., 1970. River flow forecasting through conceptual models part I. A discussion of principles. Journal of Hydrology, 10(3), 282-290.

Nelder, A.J., Mead, R., 1965. A simplex method for function minimization. Computer Journal, 7, 308-313.

Newell, R.G., Pizer, W.A., Raimi, D., 2012. Carbon Markets: Past, Present, and Future. Resources for the future discussion papers, Washington, DC, 54pp. Available online: http://www.rff.org/files/sharepoint/WorkImages/Download/RFF-DP-12-51.pdf.

Nguyen T.T.H., de Bie, C.A.J.M., Ali, A., Smaling, E.M.A., Chu, T.H., 2012. Mapping the irrigated rice cropping patterns of the Mekong delta, Vietnam, through hyper - temporal SPOT NDVI image analysis. International journal of remote sensing, 33 (2) 415-434.

Nichols, W.E., R. H. Cuenca, R.H., 1993. Evaluation of the evaporative fraction for the paramet alerization of the surface energy balance. Water Resources Research 29(11), 3681–3690, doi:10.1029/93WR01958.

Nobre, A.D., Cuartas, L.A., Hodnett, M.G., Rennó, C.D., Rodrigues, G., Siveira, A., Waterloo, M.J., Saleska, S., 2011. Height above the nearest drainage – a hydrologically relevant new terrain model. Journal of Hydrology, 404 (1–2), 13–29.

Norman, J.M., Kustas, W.P., Humes, K.S., 1995. A two-source approach for estimating soil and vegetation energy fluxes in observations of directional radiometric surface temperature. Agriculture for Meteorology, 77, 263-293.

Notter, B., Hurni, H., Wiesmann, U., Abbaspour, K.C., 2012. Modelling water provision as an ecosytem service in a large East African river basin. Hydrology and Earth System Sciences, 16, 69-86.

Nyombi, K., 2010. Understanding growth of East Africa highland banana: experiments and simulation, 198 pages. PhD Thesis, Wageningen University, Wageningen, Netherlands.

Nyombi, K., van Asten, P.J.A., Corbeels, M., Taulya, G., Leffelaar, P.A., Giller, K.E., 2010. Mineral fertilizer response and nutrient use efficiencies of East African highland banana (Musa spp., AAA-EAHB, cv. Kisansa). Field Crops Research, 117(1), 38 - 50.

Olivier F., Singels A., 2003. Water use efficiency of irrigated sugarcane as affected by row spacing and variety. Proceedings of Conference South African Sugar Technology Association, 77, 347-351.

Ozdogan, M., Gutman, G., 2008. A new methodology to map irrigated areas using multi-temporal MODIS and ancillary data: An application in the continental US. Remote Sensing of Environment, 112, 3520 - 3537.

Pande, S., van den Boom, B., Savenije, H. H. G., Gosain, A. K. 2011. Water valuation at basin scale with application to western India, Ecol. Econ., 70, 2416–2428, doi:10.1016/j.ecolecon.2011.07.025.

Pazvakawambwa, G., Van der Zaag, P., 2000. The value of irrigation water in Nyanyadzi smallholder irrigation scheme, Zimbabwe. Proceedings of the 1st

WARFSA/WaterNet Symposium 'Sustainable Use of Water Resources'. Maputo, 1-2 November.

Peng, J., Borsche, M., Liu, Y., Loew, A., 2013. How representative are instantaneous evaporative fraction measurements for daytime fluxes? Hydrology and Earth System Sciences Discussions, 10, 2015-2028, doi:10.5194/hessd-10-2015-2013.

Perry, C., Steduto, P., Allen, R.G., Burt, C.M., 2009. Increasing productivity in irrigated agriculture: agronomic constraints and hydrological realities. Agricultural Water Management 96, 151–1524.

PBWO/IUCN, 2008. Hydraulic Study of Lake Jipe, Nyumba ya Mungu Reservoir and Kirua Swamps. Pangani River Basin Flow Assessment. Pangani Basin Water Board, Moshi and IUCN Eastern and Southern Africa Regional Programme. 75 pp.

PBWO/IUCN, 2009. Hydroelectric Power Modelling Study. Pangani River Basin Flow Assessment. Pangani Basin Water Board, Moshi and IUCN Eastern and Southern Africa Regional Programme. vi+38 pp.

PBWO/IUCN, 2008. Final Project Report. Pangani Basin Water Board, Moshi and IUCN Eastern & Southern Africa Regional Program. 89 pp.

PBWO/IUCN, 2007. Pangani River System: State of the Basin Report. PBWO Moshi, Tanzania and IUCN Eastern Africa Regional Program, Nairobi, Kenya, 48pp.

PBWO/IUCN, 2006. The Hydrology of the Pangani River Basin. Report 1: Pangani River Basin Flow Assessment Initiative, Moshi, 62 pp.

PBWO/IUCN, 2009. Pangani River Basin Flow assessment. Final Project Report. Pangani Basin Water Board, Moshi and IUCN Eastern & Southern Africa Regional Programme, Nairobi. 89 pp.

Perry, C. J., 1999. The IWMI water resources paradigm – Definitions and implications. Agriculture Water Management, 40, 45-50.

Pizer, W.A., 2002. Combining price and quantity controls to mitigate global climate change. Journal of Public Economics, 85, 409-434.

Ponce-Hernandez, R., Koohafkan, P., Antoine, J, 2004. Assessing Carbon Stocks and Modelling Win-Win Scenarios of Carbon Sequestration Through Land-use Changes. Food and Agricultural Organization of the United Nations (FAO), Rome, Italy. ISBN 92-5-105168-5. 168pp.

Portmann, F.T., Siebert, S., Doll, P., 2010. MIRCA2000-Global monthly irrigated and rainfed crop areas around the year 2000: a new high-resolution data set for agricultural and hydrological modelling. Global Biogeochemical Cycles 24, GB1011.

Postel, S., 1992. Last oasis, facing water scarcity. Worldwatch environmental alert series. WW Norton and Co. New York, 272 pp.

Pouliet, D., Latifovic, R., Olthof, I., Frazer, R., 2012. Supervised classification approaches for the development of land cover in: Giri, C.P. Remote sensing and land use principles and applications, Series in remote sensing applications. CRC press, Taylor & Francis group, Pages 451, New York, pp. 174 - 185.

Prathapar, S.A. Qureshi, A.S., 1999. Mechanically reclaiming abandoned saline soils: A numerical evaluation, International Water Management Institute, Colombo, Sri Lanka.

Prince, S.D., 1991. A model of regional primary production for use with coarse-resolution satellite data. International Journal of Remote Sensing, 12, 1313-1330.

Pulido-Velazquez, M., Sahuquillo, A., Pulido-Velazquez, D., 2008. Hydro-economic river basin modelling: The application of a hollistic surface-ground water model to assess the opportunity costs of water use in Spain. Ecological Economics, 66, 51–65.

Redo, D.J., Millington, A.C., 2011. A hybrid approach to mapping land-use modification and land-cover transition from MODIS time-series data: A case study from the Bolivian seasonal tropics. Remote Sensing of Environment, 115, 353 - 372.

Reed, B.C., Brown, J.F., Van der Zee, D., Loveland, T.R., Merchant, J.W., Ohlen, D.O., 1994. Measuring Phenological variability from satellite imagery. Journal of Vegetation Science, 5, 703-714.

Reed, P.M., Hadka, D., Herman, J.D., Kasprzyk, J.R., and Kollat, J.B., 2013. Evolutionary multi-objective optimization in water resources: the past, present, and future, Advances in Water Resources, 51, 438–456.

Revelle, C., 1999. Optimization Reservoir Resources: Including a new model for reservoir reliability, John Wiley & Sons, New York, USA, 200pp.

Rijtema, P. E. Aboukhaled A., 1975. Crop water use, in: Research on crop and water use, salt affected soils and drainage in the Arab Republic of Egypt, edited by: Aboukhaled, A., Arar, A., Balba, A. M. et al., FAO, Near East Regional Office, Cairo, 5–61.

Ringler, C., Cai, X., 2006. Valuing fisheries and wetlands using integrated economic hydrologic modelling - Mekong river basin. Journal of Water Resources Planning and Management, 132, 480–487.

Rockström, J., 2003. Water for food and nature in drought-prone tropics: vapour shift in rain-fed agriculture. In Philosophical Transactions Biology, The Royal Society, London, 358, 1997 – 2009.

Rockström, J., 2000. Water resources management in smallholders farms in Eastern and Southern Africa: An overview. Physics and chemistry of the Earth, 25(3), 275-285.

Rockström, J., Falkenmark, M., Karlberg, L., Hoff, H., Rost, S., Gerten, D., 2009. Future water availability for global food production: The potential of green water for increasing resilience to global change. Water Resources Research, 45, W00A12, doi:10.1029/2007WR006767.

Rockström, J., Gordon, L., Folke, C., Falkenmark, M., Engwall, M., 1999. Linkages among water vapor flows, food production, and terrestrial ecosystem services. Conservation Ecology, 3(2):5. http://www.consecol.org/vol3/1552/art5/.

Rodríguez-Ferrero, N., 2003. Water productivity in irrigation systems. Water International, 28(3), 341 – 349, DOI: 10.1080/02508060308691708.

Roerink, G.J., Su, Z., Menenti, M., 2000. S-SEBI: A simple remote sensing algorithm to estimate the surface energy balance. Physics and Chemistry of the Earth, Part B - Hydrology, Oceans and Atmosphere, 26, 139-168.

Romaguera, M., Hoekstra, A.Y., Su, Z., Krol, M.S., Salama, M.S., 2010. Potential of Using Remote Sensing Techniques for Global Assessment of Water Footprint of Crops. Remote Sensing, 2, 1177-1196.

Romaguera, M., Kros, M.S., Salama, M.S., Hoekstra, A.Y., Su, Z., 2012. Determining irrigated areas and quantifying blue water use in Europe using remote sensing Meteosat Second Generation (MSG) products and Global Land Data Assimilation System (GLDAS) data. Photogrammetric Engineering and Remote Sensing, 78(8), 861 - 873.

Rood, S.B, Samuelson, G.M, Braatne, J.H, Gourley, C.R, Hughes, F.M.R, Mahoney, J.M., 2005. Managing river flows to restore floodplain forests. Frontiers in Ecology and the Environment, 3(4), 193–201.

Ruhoff, A.L., Paz, A.R., Collischonn, W., Aragao, L.E.O.C., Rocha, H.R., Malhi, Y.S., 2012. A MODIS-Based Energy Balance to Estimate Evapotranspiration for clear-sky days in Brazilian Tropical Savannas. Remote Sensing, 4(3), 703-725.

Saah, D., Troy, A., 2015. Developing an ecosystem service value baseline. Vietnam forests and deltas program. USAID Cooperative Agreement No. AID-486-A-12-00009. Hanoi, Vietnam.

Saatchi, S.S., Harris, N.L., Brown, S., Lefsky, M., Mitchard, E.T.A., Salas, W., Zutta, B.R., Buermann, W., Lewis, S.L., Hagen, S., Petrova, S., White, L., Silman, M., Movel, A., 2011. Benchmark map of forest carbon stocks in tropical regions across three continents, PNAS, 108(24), 9899 – 9904. www.pnas.org/cgi/doi/10.1073/ pnas.1019576108.

Sadoff, C.W., Grey, D., 2002. Beyond the river: the benefits of cooperation on international rivers, Water Policy, 4, 389-403.

Salas, J.D., Ramirez, J.A., Burlando, P., Pielke Sr, R.A., 2003. Stochastic simulation of precipitation and streamflow processes, Ch. 33, in: T.D. Potter and B.R. Colman (eds) Handbook of Weather, Climate and Water: Atmospheric Chemistry, Hydrology and Societal Impacts, Wiley & Sons, New York, pp 607-640.

Sarmett, J., Burra, R., van Klinken, R., Kelly, W., 2005. Managing water conflict through dialogue in Pangani basin, Tanzania. FAO/Netherlands International Conference on Water for food and ecosystems. The Hague, Netherlands: FAO.

Savenije, H.H.G., 1997. Determination of evaporation from a catchment water balance at a monthly time scale. Hydrology and Earth System Sciences, 1, 93-100.

Savenije, H.H.G., 2001. Equifinality, a blessing in disguise? Hydrological Processes, 15, 2835-2838.

Savenije, H.H.G., 2004. The importance of interception and why we should delete the term evapotranspiration from our vocabulary. Hydrological Processes 18, 1507–1511.

Savenije, H.H.G., 2005. Salinity and tides in alluvial estuaries. Elsevier, Amsterdam, 197pp.

Savenije, H.H.G., 2010. Topography driven conceptual modelling (FLEX-TOPO). HESS opinions. Hydrology and Earth System Sciences, 14, 2681-2692.

Scipal, K., Scheffler, C., Wagner, W., 2005. Soil moisture-runoff relation at the catchment scale as observed with coarse resolution microwave remote sensing. Hydrology and Earth System Sciences, 9, 173-183, 2005.

Schultz, G.A., 1993. Hydrological modeling based on remote sensing information, Advances in Space Research, 13(5), 149-166.

Sedjo, R.A., 2001. Forest carbon sequestration: some issues for forest investments. Discussion Paper No.: 01–34. Washington, DC: Resources for the Future.

Seyam, I.M., Hoekstra, A.Y, Savenije, H.H.G., 2003. The water value-flow concept. Physics and Chemistry of the Earth, 28, 175 – 182.

Seyfried, M.S., Wilcox, B.P., 1995. Scale and the nature of spatial variability: Field examples having implications for hydrologic modeling, Water Resources Research, 31, 173–184.

Shah, S.H.H., Vervoort, R.W., Suweis, S., Guswa, A.J., Rinaldo, A., van der Zee, S.E.A.T.M., 2011. Stochastic modelling of salt accumulation in the root zone due to capillary flux from brackish groundwater. Water Resources Research, 47, W09506, doi:10.1029/ 2010WR009790.

Shapiro, S., Wilk, M.B., 1965. An analysis of variance test for normality (complete samples), Biometrika, 52, 3 - 4, 591–611.

Shrestha, R., Tachikawa, Y., Takara, K., 2007. Selection of scale for distributed hydrological modelling in ungauged basins. IAHS Publications, 309, 290-297.

Shuttleworth, W.J., Gurney, R.J., Hsu, A.Y., Ormsby, J.P., 1989. FIFE: the variation in energy partition at surface flux sites. IAHS Publ 186, 67–74.

Sir William Halcrow & Partners, 1970. Nyumba ya Mungu resevoir and power station – Operating and maintenance instructions, London, UK.

Smakthin, V., Revenga, C., Doll, P., 2004. Taking into account environmental water requirements in global water resources assessments. Comprehensive Assessment of Water Management in Agriculture Research Report 2. Colombo, Sri Lanka: IWMI, 24p.

Smith, L. E. D., 2004. Assessment of the contribution of irrigation to poverty reduction and sustainable livelihoods, International Journal Water Resources Development, 20, 243-257.

Sotthewes, W., 2008. Forcing on the salinity distribution in the Pangani Estuary. MSc. Thesis. Delft University of Technology, Delft, the Netherlands, 82pp.

SSI, 2009. Strategies of Water for Food and Environmental security in Drought – Prone tropical and subtropical agro-ecosystems (Tanzania and South Africa). Final Report, SSI Programme. UNESCO-IHE/IWMI.

Steduto, P., 1996. Water use efficiency. In: Pereira, L.S., Feddes, L.S., Giley, R.A., Lesaffre, B. (Eds.), Sustainability of Irrigated Agriculture. NATO ASI series E: Applied Sciences. Kluwer, Dordrecht, pp. 193–209.

Steduto, P., Hsiao, T.C., Fereres, E., 2007. On the conservative behavior of biomass water productivity. Irrigation Science 25, 189–207

Steduto, P., Hsiao, T.C., Raes, D., Fereres, E., 2009. AquaCrop – The FAO crop model to simulate yield response to water: I. Concepts and underlying principles, Agronomy Journal, 101, 426–437.

Stern, N., 2007. The Economics of Climate Change: The Stern Review. Cambridge and New York: Cambridge University Press.

Stubbs, M., 2014. Conservation Reserve Program (CRP): Status and Issues. Congressional Research Service, 70-5700, R42783, 20 pages.

Su, Z., 2002. The Surface Energy Balance System (SEBS) for estimation of turbulent heat fluxes. Hydrology and Earth Systems Sciences, 6, 85-99.

Sun, Z., Wei, B., Su, W., Shen, W., Wang, C., You, D., Z. Liu, Z., 2011. Evapotranspiration estimation based on the SEBAL model in the Nansi lake wetland of China. Mathematical and Computer Modelling, 54, 1086-1092.

Teixeira, A.H.D.C., Bastiaanssen, W.G.M., Ahmed, M.D., Bos, M.G., 2009. Reviewing SEBAL input parameters for assessing evapotranspiration and water productivity for the Low-Middle Sao Franscisco River Basin, Brazil: Calibration and Validation. Agricultural and Forest Meteorology 149, 462-476.

Thompson, G.D., 1976. Water use by sugarcane. South African Sugar Journal 60, 593-600 and 627-635.

Thunnisen, H.A.M., Noordman, E., 1997. National Land Cover Database of the Netherlands: Classification Methodology and Operational Implementation. Netherlands Remote Sensing Board, Delft, the Netherlands, pp. 95.

Tilmant, A., van der Zaag, P., Fortemps, P., 2007. Modelling and analysis of collective management of water resources. Hydrology and Earth Systems Sciences, 11, 711 - 720.

Timmermans, W.J., Gieske, A.S., Kustas, W.P., Wolski, P., Arneth, A., Parodi, G.N., 2003. Determination of water and heat fluxes with MODIS imagery - Maun, Botswana. Proceeding of the International Symposium on SPIE USE V, 16, 5232 - 5255, Bellingham, Wash.

Tingting, L., Chuang, L., 2010. Study on extraction of crop information using time-series MODIS data in the Chao Phraya Basin of Thailand. Advances in Space Research, 45, 6, 775 - 784.

Townshend, J.R.G., 1981. Terrain Analysis and Remote Sensing. London. George Allen and Unwin, London.

Tsiko, C.T., Makurira, H., Gerrits, A.M.J., Savenije, H.H.G., 2012. Measuring forest floor and canopy interception in a savannah ecosystem. Physics and Chemistry of the Earth 47 - 48, 122 – 127.

Turner, D.W., Fortescue, J. A., Thomas, D.S., 2008. Environmental physiology of the bananas (Musa spp). Brazilian Journal of Plant Physiology 19, 463-484.

Turpie, J., Ngaga, Y., Karanja, F., 2003. A Preliminary Economic Assessment of Water Resources of the Pangani River Basin, Tanzania: Economic Value, Incentives for Sustainable Use and Mechanisms for Financing Management. IUCN, Nairobi, 99pp.

Uhlenbrook, S., Roser, S., Tilch, N., 2004. Hydrological process representation at the meso-scale: the potential of a distributed, conceptual catchment model. Journal of Hydrology, 291, 278–296.

UNEP, 2010. "Africa Water Atlas". Division of Early Warning and Assessment (DEWA). United Nations Environment Programme (UNEP), Nairobi, Kenya, 326pp.

USDA, 2011. Production Estimates and Crop Assessment Division (PECAD). Foreign Agricultural Service (FAS), online http://fas.usda.gov/pecad/pecad.html (accessed on 7 July 2011).

USGS, 2012. Land processes distributed active archive center (LP DAAC). Vegetation indices products, Terra and Aqua, version 5, online. https://lpdaac.usgs.gov/products.

Van Beek, L.P.H., Bierkens, M.F.P., 2009. The Global Hydrological Model PCR-GLOBWB: Conceptualization, Parameterization and Verification. Utrecht University, Faculty of Earth Sciences, Department of Physical Geography, Utrecht, the Netherlands.

Van Berkel, D.B., Verburg, P.H., 2014. Spatial quantification and valuation of cultural services in an agricultural landscape. Ecological Indicators, 37, 163 - 174.

Van Dam, J.C., Singh, R., Bessembinder, J.J.E., Leffelaar, P.A., Bastiaanssen, W.G.M., Jhorar, R.K., Kroes, J.G., Droogers, P., 2006. Assessing options to increase water use productivity in irrigated river basins using remote sensing and modelling tools. International Journal of Water Resources Development, 22, 115-133.

Van der Kwast, J., Timmermans, W., Gieske, A., Su, Z., Olioso, A., Jia, L., Elbers, J., Karssenberg, D., De Jong, S., 2009. Evaluation of the Surface Energy Balance System (SEBS) applied to ASTER imagery with flux-measurements at the SPARC 2004 site (Barrax, Spain). Hydrology and Earth System Sciences, 13, 1337-1347.

Van der Zaag, P., 2007. Asymmetry and equity in water resources management; critical institutional issues for Southern Africa. Water Resources Management, 21 (12), 1993-2004.

Van der Zaag, P., 2010. View point- Water variability, soil nutrient heterogeneity and market volatility – Why sub-Saharan Africa's Green Revolution will be location-specific and knowledge-intensive. Water Alternatives, 3(1), 154-160.

Van Vliet, J., Bregt, A. K., Hagen-Zanker, A., 2011. Revisiting Kappa to account for change in the accuracy assessment of land-use change models. Ecological Modelling, 222, 1367 - 1375.

Varlet-Grancher, C., Bonhomme, R., Charter, M., Artis, P., 1982. Efficience de la conversion de l'energie solaire par un couvert vegetal. Acta Ecologia/Ecologia Plantarum 3, 3–26.

VNFF, 2014. Payment for forest environmental services. Viet Nam Forest protection and development Fund. Newsletter No. 2 Quarter II/2014, 32pp.

Vörösmarty, C.J., Douglas, E.M., Green, P.A., Revenga, C., 2005. Geospatial indicators of emerging water stress: an application to Africa. Ambio, 34(3), 230-236.

Vörösmarty, C.J., Sahagian, D., 2000. Anthropogenic disturbance of the terrestrial water cycle. Bioscience, 50, 753-765.

Wackernagel, M., Onisto, L., Bello, P., Linares, A.C., Falfán, I.S.L., García, J.M., Guerrero, A.I.S., 1999. National natural capital accounting with the ecological footprint concept (Analysis). Ecological Economics, 29, 375 – 390.

Waclawovsky, A.J., Sato, P.M., Lembke, C.G., Moore, P.H., Souza, G.M., 2010. Sugarcane for bio-energy production: an assessment of yield and regulation of sucrose content. Plant Biotechnology Journal 8, 263–276.

Wallace, S., Fleten, S., 2003. Stochastic programming models in energy. In Stochastic programming, Handbooks in operations research and management science, edited by A. Ruszczynski and A. Shapiro, vol. 10, 637 - 677, North-Holland.

Ward, F.A., Hurd, B.H., Rahmani, T., Gollehon, N., 2006. Economic impacts of federal policy responses to drought in the Rio Grande Basin, Water Resource Research, 42, W03420, doi:10.1029/2005WR004427.

Wardlow, B.D., Egbert, S.L., 2008. Large area crop mapping using time-series MODIS 250 m NDVI data: an assessment for the U.S. Central Great Plains. Remote Sensing of Environment, 112, 1096-1116.

WCD, 2000. Dams and Development - A New Framework for Decision-Making. The Report of the World Commission on Dams, Earthscan publication, London, 322pp.

Wiegand, C.L., Richardson, A.J., Escobar, D.E., Gerberman, A.H., 1991. Vegetation indices in crop assessments. Remote Sensing of Environment 35, 105–119.

Wilson, E.M., 1983. Engineering Hydrology (3rd Ed.), Macmillan Education Ltd, London, UK.

Winsemius, H.C., 2009. Satellite data as complementary information for hydrological modelling. PhD dissertation. Delft University of Technology, Delft, the Netherlands.

Winsemius, H.C., Savenije, H.H.G., Bastiaanssen, W.G.M., 2008. Constraining model parameters on remotely sensed evaporation: justification for distribution in ungauged basins? Hydrology and Earth System Sciences, 12, 1403-1413.

Winsemius, H.C., Savenije, H.H.G., Gerrits, A.M.J., Zapreeva, E.A., Klees, R., 2006. Comparison of two model approaches in the Zambezi river basin with regard to model reliability and identifiability. Hydrology and Earth System Sciences, 10, 339-352.

World Bank, 2006. Water Resources Assistance Strategy. Improving Water Security for Sustaining Livelihoods and Growth. United Republic of Tanzania, Report No. 35327 - TZ. World Bank.

Yan, N., Wu, B., 2014. Integrated spatial–temporal analysis of crop water productivity of winter wheat in Hai Basin. Agricultural Water Management, 133, 24-33.

Yokwe, S., 2009. Water productivity in smallholder irrigation schemes in South Africa, Agricultural Water Management, 96, 1223-1228.

Young, R.A., 2005. Determining the Economic Value of Water: Concepts and methods. Resource for the Future, Washington, 374pp.

Zhang, G.P., Savenije, H.H.G., 2005. Rainfall-runoff modelling in a catchment with a complex groundwater flow system: application of the Representative Elementary Watershed (REW) approach. Hydrology and Earth System Sciences, 9, 243 - 261.

Zhang, S., Zhao, H., Lei, H., Shao, H., Liu, T., 2015. Winter Wheat Water Productivity Evaluated by the Developed Remote Sensing Evapotranspiration Model in Hebei Plain, China. The Scientific World Journal, vol. 2015, Article ID 384086, 10 pp, doi:10.1155/2015/384086.

Zhang, X., Pei, D., Hu, C., 2003. Conserving groundwater for irrigation in the North China Plain. Irrigation Sciences, 21, 159–166.

Zhang, X., Rui, S., Bing, Z., Qingxi, T., 2008. Land cover classification of the North China Plain using MODIS-EVI time series. ISPRS Journal of Photogrammetry & Remote Sensing, 63, 476-484.

Zhao, M., Heinsch, F.A., Nemani, R., Running, S.W., 2005. Improvements of the MODIS terrestrial gross and net primary production global data set. Remote Sensing of Environment, 95, 164-176.

Zhao, M., Running, S.W., Nemani, R.R., 2006. Sensitivity of Moderate Resolution Imaging Spectroradiometer (MODIS) terrestrial primary production to the accuracy of meteorological reanalyses. Journal of Geophysical Research, Vol. 111, No. G1, G01002.

Zwart, S.J., Bastiaanssen, W.G.M., 2004. Review of measured crop water productivity values for irrigated wheat, rice, cotton and maize. Agricultural Water Management 69, 115-133.

Zwart, S.J., Bastiaanssen, W.G.M., 2007. SEBAL for detecting spatial variation of water productivity and scope for improvement in eight irrigated wheat systems. Agriculture Water Management, 89, 287-296.

Zwart, S.J., Bastiaanssen, W.G.M., de Fraiture, C., Molden, D.J., 2010. WATPRO: A remote sensing based model for mapping water productivity of wheat. Agricultural water management, 97, 1628-1636.

Zwart, S.J. and Leclert, L.M.C., 2010. A remote sensing based irrigation performance assessment: a case study of the Office du Niger in Mali. Irrigation Science, 28, 371-385.

SAMENVATTING

Het integraal waterbeheer concept (IWRM) heeft als doel om alle relevante elementen van water te integreren op een omvattende en holistische manier. Een integraal waterbeheerplan moet oog hebben voor het gecombineerde beheer van blauw en groen water in een stroomgebied en hun ruimtelijke en temporele verdeling. Groen en blauw water volgen verschillende routes en worden geassocieerd met verschillende watergebruikspraktijken. In sub-Sahara Afrika worden bovenstrooms gelegen landschappen gedomineerd door regen-afhankelijke en supplementair-geïrrigeerde gewassen, welke voornamelijk aangewezen zijn op groen water. In stroomafwaarts gelegen gebieden is het gebruik van blauw water beperkt tot de nabijheid van rivieren, voornamelijk voor waterkracht en voor ecosystemen. In de loop van de tijd, en als gevolg van de bevolkingsgroei en de toegenomen vraag naar voedsel en energie, is de vraag naar zowel groen als blauw water toegenomen. Het toegenomen gebruik van groen water in bovenstroomse delen van stroomgebieden heeft vaak geleid tot dalende beschikbaarheid van blauw water in benedenstroomse delen. De klassieke aanpak van waterbeheer richt zich vaak slechts op blauw water (rivierafvoer). Dit kan worden toegeschreven aan beperkte informatie over de temporele en ruimtelijke verdeling van het groene water (bodemvocht) in een stroomgebied. Uiteraard heeft dit de ontwikkeling van omvattende en duurzame IWRM plannen in dergelijke stroomgebieden belemmerd. Om afhankelijkheden tussen bovenstroomse en benedenstroomse gebieden te beoordelen, en randvoorwaarden te identificeren voor een optimaal waterbeheersplan op stroomgebiedsniveau, is daarom een geïntegreerd analyse-systeem voor het hele stroomgebied nodig, waarin zowel het groene als het blauwe water wordt vervat.

Voor het beheer van de onderlinge afhankelijkheden in een stroomgebied - in het bijzonder in geval van waterschaarste – zijn beschikbare kennis en gegevens van fundamenteel belang. Dit proefschrift heeft verschillende benaderingen - waarvan sommigen worden beschouwd als innovaties – toegepast om lokaal gevalideerde informatie te generen voor een heterogeen, sterk benut en gegevens-schaars stroomgebied in Afrika, namelijk de Pangani. Door een nauwkeurige evaluatie van (i) de beschikbaarheid van water, (ii) het gebruik van water, (iii) de productiviteit van water, en (iv) de waarde van water, konden de onderlinge afhankelijkheden in het stroomgebied, alsmede de kwantificering van compromissen en synergiën tussen de verschillende gebruikers van groen en blauw water, geïdentificeerd worden.

Het stroomgebied van de Upper Pangani kan als gesloten worden beschouwd vanwege het intensieve gebruik van water, vooral in de landbouw. De vele irrigatiesystemen ontwikkeld door kleine boeren bestaan uit complexe en ingewikkelde netwerken van aarden kanalen die supplementaire irrigatie leveren aan gewassen die grotendeels regenafhankelijk zijn, waarbij regenwater (groen water) en rivierwater (blauw water) gecombineerd worden. Er is zeer weinig officiële informatie over het watergebruik en de waterproductiviteit van deze irrigatiesystemen. Het toenemende watergebruik voor irrigatie in bovenstroomse gebieden heeft geleid tot externaliteiten en water-gerelateerde conflicten tussen de verschillende gebruikers in het bekken. Met de tijd zijn ook ecosystemen negatief beïnvloed aangezien de meeste zijrivieren niet langer water voeren gedurende de droge tijd.

In semi-aride gebieden als het Upper Pangani stroomgebied vormt de verdamping de grootste component van de hydrologische cyclus, terwijl rivierafvoer nauwelijks meer is dan 10%. De verdampingsterm is een functie van bodembedekking en landbeheer. Informatie over het ruimtelijke landgebruik en de bodembedekking (LULC) is nodig om (i) het groen watergebruik per LULC vast te stellen, en (ii) die parameters voor een hydrologisch model te karakteriseren welke het groene water met het blauwe water verbinden. Door middel van remote sensing werden zestien verschillende LULC klassen geïdentificeerd en ingedeeld aan de hand van hun unieke temporele fenologische signatuur. De methode gebruikte vrij beschikbare satellietgegevens van vegetatie van de Moderate-resolution Imaging Spectroradiometer (MODIS). De gegevens hebben een resolutie van 8 dagen (temporeel) en 250 m (ruimtelijk), en betreffen de hydrologische jaren van 2009 tot 2010. Automatische en begeleide clustering technieken werden gebruikt om verschillende soorten LULC te identificeren op basis van grond observaties uit het stroomgebied tijdens de twee regen seizoenen (de korte en lange regens). De multi-temporele MODIS data en de lange tijdreeks zorgde voor een juiste timing van de veranderingsmomenten in de vegetatie groei. De algemene classificatie nauwkeurigheid was 85%, met een samensteller nauwkeurigheid van 83% en een gebruiker nauwkeurigheid van 86% (op 98% betrouwbaarheidsniveau). De individuele klassen behaalden relatief goede nauwkeurigheden, groter dan 70%, met uitzondering van braakliggende gronden. De kleinere LULC klassen haalden lagere nauwkeurigheden. Deze onzekerheid werd toegeschreven aan de matige raster resolutie van MODIS (250 m). De onjuistheden werden gecorrigeerd met behulp van de Kappa statistiek (K). De LULC klassen waren consistent met de FAO-SYS landgeschiktheidsclassificatie. Lokale databases van kleinschalige landbouw en grootschalige irrigatie plantages (suikerriet) werden gebruikt voor additionele controles, en er werden nauwe overeenkomsten gevonden (74% en 95%, respectievelijk), met een vrij goede geografische spreiding.

Het nauwkeurige schatten van de werkelijke verdamping (ET) voor de 16 verschillende LULC klassen in een regio waar gegevens schaars zijn is een uitdaging. Deze studie gebruikte de MODIS satellietgegevens en de Surface Energy Balance Algorithm of Land (SEBAL) om de werkelijke ET te schatten op basis van 138 beelden, met 250-m en 8-daagse resolutie voor de periode 2008-2010. Er was een goede overeenkomst tussen de SEBAL ET en verschillende validaties. De geschatte ET (open water) voor Nyumba ya Mungu (NYM) reservoir had een goede correlatie met de gemeten pan verdamping ($R^2 = 0,91$; Root Mean Square Error (RMSE) van minder dan 5%). Een absolute relatieve fout van 2% werd gevonden op basis van de gemiddelde jaarlijkse waterbalans van het reservoir. De geschatte ET voor de landbouw gebruiksklassen waren consistent met de seizoensgebonden variabiliteit van de gewas coëfficiënt (K_c) op basis van de Penman-Monteith vergelijking. De ET ramingen voor bergachtige gebieden werden significant onderdrukt op grotere hoogte (boven 2300m), wat overeenkomt met de verminderde potentiële verdamping in die gebieden. De ET schattingen waren vergelijkbaar met de globale MODIS 16 ET data set wat betreft de variantie van de gegevens (significant met 95% betrouwbaarheid), maar niet voor de gemiddelde waardes. Deze significantie biedt optimisme maar tegelijkertijd ook voorzichtigheid bij het gebruik van de vrij beschikbare globale ET datasets, omdat deze niet lokaal zijn gevalideerd.

Een belangrijke beperking in het afleiden van remote-sensed ET, speciaal voor landgebruikstypes op hoger gelegen gebieden in de humide tot sub-humide tropen, is aanhoudende bewolking. De pixels met bewolking moesten worden gecorrigeerd door interpolatie op basis van de volgende en/of voorgaande beelden. Hoewel er gebruik is gemaakt van de multispectrale reeksen van de MODIS beelden, kan deze vulprocedure nog steeds onzekerheden in de uiteindelijke resultaten introduceren. Voor het gehele stroomgebied was de geschatte ET 94% van de totale regenval, wat resulteerde in een afvoer bij de uitlaat van het stroomgebied dat 12% verschilde van de gemeten afvoer. De afwijking (12%) viel binnen de onzekerheidsmarge (13%) bij 95% betrouwbaarheid. De waterbalans analyse toonde duidelijk aan dat het stroomgebied snel aan het sluiten is. Daarom is het belangrijk en tijdig om de waterproductiviteit te verhogen door middel van verbeterde waterefficiëntie en water reallocatie in het Upper Pangani stroomgebied.

Het kwantificeren van het hydrologische verband tussen het ruimtelijk gebruik van groen (verdamping) en blauw water (rivier afvoer) is van essentieel belang voor de beoordeling van onderlinge afhankelijkheden op stroomgebiedsniveau, maar is een uitdaging. Fysiek-gebaseerde ruimtelijke modellen worden vaak gebruikt. Maar deze modellen vereisen enorme hoeveelheden gegevens, wat kan leiden tot equifinaliteit, wat dergelijke modellen minder geschikt maakt voor scenarioanalyses. Bovendien richten deze modellen zich meestal op natuurlijke processen en houden geen rekening met antropogene invloeden. Deze studie heeft een innovatieve methode gebuikt om de blauwe en groene waterstromen te kwantificeren. De methode gebruikt ET en bodemvocht data die afgeleid zijn van remote-sensing als input in het Spatial Tools for River basin Environmental Analysis and Management (STREAM) model. Om de wijdverbreide irrigatiewater onttrekkingen in het model mee te nemen werd een extra blauw water component (Q_b) opgenomen in het STREAM model om het irrigatiewater gebruik te kwantificeren. Om model parameter identificatie en kalibratie te vergemakkelijken werden twee hydrologische landschappen (moerassen en heuvels) onderscheiden op basis van veldgegevens en topografische kaarten. Het model werd gekalibreerd met afvoergegevens van vijf meetstations, met goede resultaten vooral wat betreft de simulatie van episodes van laagwater. De natuurlijke logaritme Nash-Sutcliffe Efficiency (Ens_ln) van de afvoer waren groter dan 0,6 zowel voor de kalibratie en validatie periodes. De Ens_ln coëfficiënt van de afvoer aan het eind van het stroomgebied was nog hoger (0,90). De enige uitdaging met het gebruik van remote sensing data (met een 8-daagse interval) als input in hydrologische modellen betreffen processen met tijdschalen korter dan 8 dagen, zoals interceptie. Dergelijke hydrologische processen moeten worden berekend buiten het model om, en dragen zo bij aan extra onzekerheden.

Tijdens laagwater verbruikte Q_b bijna 50% van het rivierwater in het Upper Pangani bekken. Q_b voor irrigatie was vergelijkbaar met de veld-gebaseerde netto irrigatie schattingen, met een afwijking van minder dan 20%. Een aantal waterbeheerscenario's met betrekking tot waterbesparing en de effecten van het toegenomen watergebruik werden verkend. Het gewijzigde STREAM model kan gerepliceerd worden in andere landschappen met complexe interacties tussen groen en blauw water. De flexibiliteit van het model biedt de mogelijkheid voor een doorlopende verbetering ervan zodra meer gegevens beschikbaar komen. De output van het model, met name de informatie over groene en blauwe waterstromen, werd gebruikt als input voor de analyse van de waterproductiviteit.

Hoewel de waterproductiviteit een belangrijke indicator is in het waterbeheer van stroomgebieden is het niet direct beschikbaar, in het bijzonder voor natuurlijke landschappen. De maatregelen om de waterproductiviteit te verbeteren verschillen ook per stroomgebied. Deze studie heeft de waterproductiviteit in de Upper Pangani berekend met behulp van een combinatie van remote-sensing modellen. De modellen waren gebaseerd op de Monteith's droge-stof productie methode om de bovengrondse biomassa productie in de landbouw en in natuurlijke landschappen te schatten. Het SEBAL algoritme werd gebruikt voor het berekenen van de biomassa productie op basis van MODIS beelden. De ruimtelijke informatie van biomassa productie werd vervolgens omgezet in gewasopbrengst en hoeveelheid vastgelegde koolstof. Deze werden vervolgens omgezet in bruto opbrengsten met behulp van marktprijzen. Deze studie incorporeerde het bruto rendement van carbon credits en van andere ecosysteemdiensten in het concept van de economische waterproductiviteit (EWP). De EWP liet de niveaus van het watergebruik zien; indien geformuleerd als productiefuncties toont het de mogelijkheden voor verbeteringen en kan het een trade-off analyse maken op stroomgebiedsniveau. De biofysische productiviteit (biomassa en gewasopbrengst) en water opbrengst maakte ook inzichtelijk wat de water waarde is die een samenleving toekent aan bepaalde natuurlijke landgebruiken.

Geïrrigeerde suikerriet en rijst behaalden de hoogste waterproductiviteit, zowel biofysisch als economisch - ruim binnen de waardes die in de literatuur genoemd worden. De waterproductiviteit van regen-afhankelijke en supplementair-geïrrigeerde bananen en maïs vertoonden een grote ruimtelijke variabiliteit en was significant lager dan potentieel. De supplementair-geïrrigeerde gewassen die groen en blauw water combineren behaalden echter een hogere economische productiviteit van blauw water dan volledig geïrrigeerde gewassen. In situaties van waterschaarste is het daarom verstandig om water te alloceren aan supplementair-geïrrigeerde gewassen in plaats van aan volledig geïrrigeerd gewassen. Dit proefschrift ontwikkelde expliciete analytische relaties tussen biomassa productie en ET voor geïrrigeerde, regen-afhankelijke en natuurlijke landschappen in het Pangani stroomgebied. Deze relaties, geformuleerd als productiefuncties, toonden het potentieel van het verbeteren van de productiviteit van de regen-afhankelijke en supplementair-geïrrigeerde landbouw in het stroomgebied. De frequentieverdeling van de biomassa productie op pixel-niveau gaf aanvullende bewijs voor de verbetering van de waterproductiviteit.

Een geïntegreerde hydro-economisch model (IHEM) werd ontwikkeld om groen en blauw water te integreren voor de meervoudige doelen analyse van het watergebruik in het gehele Pangani stroomgebied. De IHEM, die gericht is om het blauw watergebruik te optimaliseren, werd innovatief geformuleerd om rekening te houden met de volledige waterbalans. Dit is gedaan door het opnemen van groen water door middel van hun productie functies in de Upper Pangani. De analyse richt zich op drie primaire doel functies: i) waterkracht productie, ii) volledig geïrrigeerde landbouw, waarbij de water behoefte geheel door blauw water wordt vervuld, en iii) supplementaire irrigatie, waar de waterbehoefte van de gewassen wordt vervuld door zowel groene als blauw water. De analyse beschouwde ook vijf socio-ecologische doelstellingen die waren geïnformeerd door de belangrijkste stakeholders en door vakkennis. De resultaten toonden aan dat de (supplementair- en volledig geïrrigeerde) landbouw een relatief hoge waterproductiviteit behaalt wat concurreert met waterkracht, stedelijke watergebruik en ecosystemen. Betrouwbare energie (90% betrouwbaar) heeft een constante gematigde rivierafvoer nodig gedurende het gehele jaar; deze concurreert met het milieu dat zowel hoge als lage rivierafvoeren vereist, afhankelijk van de seizoenen. Deze studie vond dat de verbetering van regenafhankelijke maïs door supplementaire irrigatie een iets hogere marginale water waarde heeft dan de volledige geïrrigeerde suikerriet. Om duurzaamheid van het stroomgebied te bereiken, moet het watergebruik van de landbouw worden afgewogen tegen andere economische, sociale en ecologische watergebruiken. Omdat de waterbehoefte voor waterkracht grotendeels nonconsumptief is, kan, althans in theorie, waterkracht productie seizoens-afhankelijk gevarieerd worden, zodat het synchroniseert met de water behoefte van het ecosysteem.

Het IHEM model genereerde de blauwe waterbalans van het Lower Pangani stroomgebied en toonde aan dat de Upper Pangani 82% bijdraagt van de totale blauw water hoeveelheid. Verdamping van het NyM reservoir bedraagt ongeveer 28% van de totale instroom in het reservoir. Het gebruik van water in het Kirua moeras, hoewel beperkt vanwege de regulering van de rivierafvoer door het NyM reservoir, vertegenwoordigt een waarde van 8 miljoen US$ per jaar aan potentiële waterkracht inkomsten. De studie toonde aan dat de minimale water behoefte van het ecosysteem in het Pangani estuarium gegarandeerd wordt door de waterbehoefte van de twee waterkrachtcentrales niet ver bovenstrooms daarvan. Verder wordt de behoefte van een seizoensgebonden hoger debiet in de monding momenteel ondersteund door ongereguleerde afvoeren van de Mkomazi en Luengera zijrivieren. De scenario-analyses toonden verschillende niveaus van trade-offs tussen concurrerende watergebruikers. Elke maatregel die de instroom naar het reservoir verhoogt, of de waterbehoefte benedenstroom verkleint, resulteert in een operationeel beheer dat reservoir verdamping minimaliseert en zorgt voor een natuurlijker rivierafvoer stroomafwaarts. Investeringen in interventies die de niet-productieve verdamping van bodemvocht in de geïrrigeerde gemengde gewassen in bovenstroomse gebieden verminderen resulteerde in een verhoogde blauw water instroom in NyM reservoir dat de waterkracht productie zou doen stijgen met 2 miljoen US$ per jaar. Dit komt overeen met 33 US$ ha^{-1} jr^{-1}, dat beschikbaar is voor investeringen in bodem en water conservering, mogelijk in de vorm van betaling voor milieudiensten (PES). De toename van de omzet komt bovenop de niet-gekwantificeerde additionele ecosysteemdiensten die zouden voortvloeien uit toegenomen benedenstroomse rivierafvoeren.

Hoewel dit onderzoek duidelijk de voordelen van geïntegreerd hydro-economisch modelleren kon aantonen door het opnemen van groen watergebruik in bovenstroomse gebieden en blauw watergebruik benedenstrooms, bleek het afleiden van een nauwkeurige water waarde voor ecosysteemdiensten, met name voor moerassen, een uitdaging. De waarde van ecosysteemdiensten kunnen worden geïncorporeerd in de niet-economische productiefuncties (gebruikt als beperkingen in ons model) om zodoende een breder scala aan opties en trade-offs voor belanghebbenden en besluitvormers te genereren.

About the Author

Jeremiah Kipkulei Kiptala, born in 1976 in Baringo County, Rift Valley, Kenya, obtained his Bachelor of Science (BSc) degree in Civil Engineering in 2000 from the University of Nairobi, Kenya. He was later engaged as a graduate engineer in a civil engineering consulting firm. In 2003, he joined the National Water Conservation and Pipelice Corporation (NWCPC) as an assistant engineer, urban water supply. In 2004, he was appointed the resident engineer by NWCPC for the larger northern and arid region of Kenya to be incharge of water resource development works. He was involved in the construction of small dams and water pans. He was also involved in the drilling of boreholes, water supply, flood protection and diversion works in the lower Tana River Basin, Kenya.

In 2006, he joined UNESCO-IHE, Delft, the Netherlands, to pursue the Master of Science (MSc) Degree in Water Management specialized in Water Resources Management. His MSc thesis was entitled 'Intersectoral Water Allocation in the Tana River Basin'. Jeremiah graduated in 2008 (with distinction). Thereafter, he was appointed by NWCPC as the project engineer for the Maruba Dam, a large dam construction project in Eastern Kenya. The project was successfully completed and commissioned in May 2010 to serve Machakos Township. Later that year, he received a scholarship from UNESCO-IHE to pursue a PhD in the field of water resources management under the SSI project in Eastern Africa. In 2011, he joined the department of Civil, Construction and Environmental Engineering, Jomo Kenyatta University of Agriculture and Technology (JKUAT), Kenya as a lecturer in water resources engineering.

During his PhD study, he specialized in remote sensing, hydrological and river system modelling and ecosystem services valuation. He has supervised several MSc research theses (both at UNESCO-IHE and JKUAT) and BSc Civil Eng. research projects at JKUAT. He has also consulted for the International Water Management Institute (IWMI) on various research activities in the Tana River basin, Kenya. He has several publications in international journals and conferences.

Currently, he is a lecturer and the Chairman of the Department of Civil, Construction and Environmental Engineering and the executive board member of Water Research and Resource Centre (WARREC) of JKUAT, Nairobi, Kenya. He is also a registered and practising professional civil engineer (Engineers Board of Kenya) and a corporate member of the Institute of Engineers of Kenya. He has consulted with various consultancy firms in Kenya.

Printed and bound by CPI Group (UK) Ltd, Croydon, CR0 4YY

18/10/2024

01776210-0004